UNDERSTANDING CANCER FROM A SYSTEMS BIOLOGY POINT OF VIEW

UNDERSTANDING CANCER FROM A SYSTEMS BIOLOGY POINT OF VIEW

From Observation to Theory and Back

IRINA KAREVA

ELSEVIER

ACADEMIC PRESS
An imprint of Elsevier

Academic Press is an imprint of Elsevier
125 London Wall, London EC2Y 5AS, United Kingdom
525 B Street, Suite 1800, San Diego, CA 92101-4495, United States
50 Hampshire Street, 5th Floor, Cambridge, MA 02139, United States
The Boulevard, Langford Lane, Kidlington, Oxford OX5 1GB, United Kingdom

Library of Congress Cataloging-in-Publication Data
A catalog record for this book is available from the Library of Congress

British Library Cataloguing-in-Publication Data
A catalogue record for this book is available from the British Library

ISBN: 978-0-12-813673-7

Publisher: Mica Haley
Acquisition Editor: Rafael Teixeira
Editorial Project Manager: Lisa Eppich
Production Project Manager: Kiruthika Govindaraju
Designer: Limbert Matthew

Typeset by TNQ Books and Journals

"To my family"

CONTENTS

CHAPTER 1

Introduction

Contents

Research is driven by the interplay of theory and experiment. Experiments give rise to theories, which we test using experiments, which drive theory, until we find answers to questions that interest us. After all, research is first and foremost about satisfying scientific curiosity.

Some experiments are easy to run. Most are not. Experiments require planning, and for most biological questions, expensive equipment, laboratory space, and abundant resources. Even more importantly, investigating any biological system means trying to unravel the interplay of numerous factors, the influence of which on the final outcome is not always easy to tease out. That is where mathematical modeling can come in very handy.

The schematic approach for using mathematical modeling for testing hypotheses about biological mechanisms can be summarized as follows:

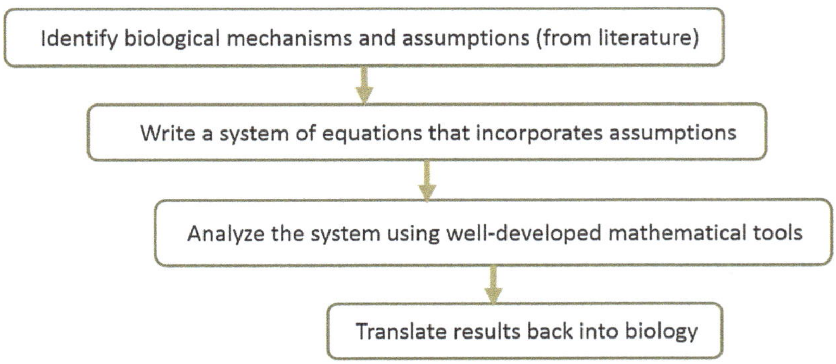

In a way, mathematical modeling is like Rosetta Stone, a way to translate back and forth between mathematics and biology. As modelers, we try to (1) understand a biological mechanism, (2) tease out key players that drive the dynamics, (3) formulate assumptions about how these key players interact with each other and their environment, (4) translate these assumptions into equations, (5) analyze them using a very well-developed mathematical

Understanding Cancer From a Systems Biology Point of View
ISBN 978-0-12-813673-7
http://dx.doi.org/10.1016/B978-0-12-813673-7.00001-X

apparatus, and (6) translate the results back into the language of biology. If our predictions are consistent with experimental observations, then maybe we got it right, and we can now try to make further predictions about what will happen if we change this or that aspect of the system. Conducting early in silico experiments can save thousands of dollars when planning further experiments, advancing a "fiscally responsible" way of conducting research.

And what if the predictions made by the model are not consistent with experimental observations? That may be even more interesting. This means that some of our assumptions are incorrect, and thus our understanding of biology is incomplete. The good news is that since we built the model, we control all the assumptions, and so we can go through them one by one and identify which one is causing the discrepancy. Chances are we just identified a gap in knowledge, which we can now try to close using both theory and experiment.

However, all of this matters only when there is a question. A question that drives your curiosity, a question that bugs you like a pebble in your shoe. The question driving the work that led to this book stems from the following observation: ***sometimes after removal of primary tumors, metastases start growing faster***. This has been observed particularly in colorectal cancer but in other cancer types as well. And the question is **Why?** What is going on?

It is possible that removing a primary tumor can result in some kind of malignant transformation, causing secondary tumors to form and rapidly start growing, but it is not very likely. Although some aggressive tumor types do exist, tumors typically take years and sometimes decades to form and grow. It is more likely that secondary tumors were there all along, either kept at bay or "sleeping," and something either about the process of primary tumor removal, such as surgery, or the changes that happen in the body after the removal of the primary tumor "woke them up." Cancers that are present but do not grow are known as dormant tumors, cancer without disease. Trying to understand mechanisms that underlie the state of dormancy might hold a key to answering the question underlying this book.

However, before we dive into that, it is important to gain some understanding about the basic mechanisms of mathematical modeling.

A PRIMER ON MATHEMATICAL MODELING: PREDATOR–PREY

A key aspect of understanding dynamical systems and writing mathematical models lies in thinking not of what things are but of what they do. This translates into thinking about relationships between individuals (people, animals, or cells), and how they interact with each other and their environment.

For instance, predator–prey type models have been developed extensively in the literature, starting from the classic Lotka–Volterra models that describe the dynamics of two populations, where the growth of one is dependent on consuming the other. These models were used to explain the dynamics of natural populations of lynx and snowshoe hare in Canada, for instance. However, predator–prey type models are also the classic framework for describing tumor–immune interactions, where tumor is the "prey" and immune system is "predator." We will use this framework as an example for describing how to write a system of ordinary differential equations, which will be used in later chapters.

Equations describing tumor growth typically have the following components:

- Growth terms, often logistic or "Gompertz," of which logistic form is a special case
  - The carrying capacity (maximum population size that can be supported indefinitely) may remain fixed or can be dynamic, depending on the type of question the authors may be asking

  Death terms:
  - Death due to apoptosis or preprogrammed natural death (this can be included in logistic growth or as a separate term)
  - Death due to predation by the immune system
  - Death due to the activity of a drug or chemical agent

Equations describing the dynamics of the immune cells typically focus on CD8+T cells (white blood cells that have originated in the thymus and that express the CD8 proteins on cell surface) as the most efficient cytotoxic predators; however, they can be expanded to include antigen presenting cells (APCs), natural killer (NK) cells, etc., depending on the questions and the level of detail that is necessary to answer them. Notably, increasing the level of detail is *not* necessarily beneficial and can make it more difficult to discern the effects of the processes investigated. Therefore, the level of detail should be driven by the question, and only cell types and agents that quantitatively affect the dynamics, directly or indirectly, should be taken into account.

Equations describing the dynamics of immune cells typically have the following components:

Growth terms:

- Natural inflow (population growth) rates to describe baseline levels of the immune cells.

- Additional inflow of immune cells due to the presence of the tumor. Since the immune cells react to the antigens/debris of the previously killed tumor cells, this term is proportional to the number of previously killed tumor cells. This is an important difference of tumor–immune models as compared to classic predator–prey type models, where increase of the predator would be directly proportional to how much prey it has consumed, as opposed to how many "carcasses" the predator has encountered.
- In some variations, there may be other factors contributing to the ability of the predator (immune system) to grow. For instance, in Chapter 3 on tumor–immune interactions, we looked at the situation where both tumor and immune cells compete for glucose, a shared resource that is necessary for both populations to grow. Therefore, in this model, the term describing immune cell growth was also dependent on the level of available glucose.

Death terms:

- There often exists a natural death term for the tumor-specific immune cells to indicate decrease in immune cell levels following decrease in tumor levels.
- There can also be a treatment-induced death term, which accounts for immune cell sensitivity to many cytotoxic drugs.

Once written, a system of differential equations is coded into a numerical solver, such as Matlab, and investigated numerically (in Matlab it is typically done using functions ode45 or ode23). To conduct a numerical investigation of a dynamical system, we identify parameters of interest for the question (this can be done based on understanding of the biology or through conducting sensitivity analysis, which will be discussed in later chapters), vary them, and create plots of how predicted population sizes change over time. This approach provides insights into what processes may be driving the dynamics of the investigated system, how different components of the system affect each other, and what may or may not be reasonable to try and reproduce in an experimental setting. Numerous examples will be given later in the book showing how this process is done.

ESTIMATING PARAMETERS

Many models appear to be overparameterized. As a rule, there should be as many parameters as are necessary for the units to balance out on both sides of the equation. That is, if we are tracking rate of change of population size over time on the left-hand side of the equation, we might have to introduce scaling parameters in the right-hand side terms to ensure balanced units.

Estimating parameter values for simulations may prove challenging because it may not always be possible to find experiments that will allow discerning exact values. Furthermore, in experiments, we may find that several parameters are combined into one, and discerning individual values for each may not always be possible or useful.

However, even in the absence of exact parameter values, it is important to be able to estimate the order of magnitude of the parameter value due to potential differences in time scales. If one of the described processes occurs qualitatively faster or slower than others, then failing to take this into account would attribute importance to a process that may not actually be affecting the dynamics, thus leading to incorrect or misleading predictions.

If one has access to experimental data, it may be possible to find parameters that allow reproducing the data using parameter fitting approaches such as least squares, Monte Carlo simulations, or built-in Matlab functions such as fmincon and fminsearch. This is also one of the best ways to validate the model before proceeding with further investigation.

When using published data to estimate parameters, it is often difficult to obtain exact values since raw data values are rarely published. However, data values can be read off of published graphs using tools such as WebPlotDigitizer. Otherwise, in the absence of data, we rely on logic and common sense and do the best we can to get the quantitative behaviors right.

With the above as background, let us dive into the first part of the quest for the answer: trying to understand mechanisms that can lead to dormant tumors.

CHAPTER 2

Tumor Dormancy

Contents

INTRODUCTION

Tumors do not just appear. Autopsy studies of adults who died of trauma have revealed a surprisingly high number of microscopic clinically unapparent tumors. For instance, microtumors of the breast were found in 39% of 39- to 50-year-old women, while only 1% of them would have received breast cancer diagnosis in their lifetime. Similar results were obtained for prostate cancer—46% of 60- to 70-year-old men had microtumors of the prostate but only 1%−1.5% of them would receive a diagnosis of prostate cancer in their lifetime. The most striking example, however, was that of thyroid cancer—98% of 50- to 70-year-old people had microtumors of the thyroid, and only 0.1% would ever develop the disease [1−4] (Table 2.1).

This phenomenon has become known as tumor dormancy or "cancer without disease" [5]. It can manifest itself not only in preclinical disease but also in long latency periods between removal of primary tumors and appearance of metastases (secondary or metastatic dormancy) [6−9]. It can even appear in some unfortunate organ transplant recipients [10−12], who can develop disease after microscopic tumors were transmitted from donors who passed pretransplantation screening and have been free from cancer for over 10 years.

Understanding Cancer From a Systems Biology Point of View
ISBN 978-0-12-813673-7
http://dx.doi.org/10.1016/B978-0-12-813673-7.00002-1

Table 2.1 Findings from autopsy studies of people who died of trauma, revealing vast differences between microtumor incidence and lifetime diagnosis

Cancer type	Age group	Microtumor incidence	Lifetime diagnosis
Breast carcinoma in situ	39- to 50-year-old women	**39%**	1%
Prostate cancer	60- to 70-year-old men	**46%**	1%–1.5%
Thyroid cancer	50- to 70-year-old people	**98%**	0.1%

Within dormant tumors, two general mechanisms underlying the dynamics are recognized: (1) when tumor cell population as a whole remains at a limited carrying capacity through balance of proliferation and apoptosis (tumor mass dormancy) and (2) solitary cell dormancy, when individual cells enter a state of quiescence.

A number of mechanisms have been proposed to explain the phenomenon of tumor dormancy. These include cell cycle arrest, when cells become incapable of dividing further [13,14]; tumor management by the immune system [15,16], when cytotoxic lymphocytes effectively recognize and eliminate sufficient numbers of cancer cells, keeping the tumor at a constant size; decreased relative fitness of tumor cells relative to normal cells in younger tissues [17]; inability to secure blood supply, which prevents access to oxygen and nutrients [18,19], among others. Two of these, namely tumor–immune interactions and blood vessel formation mechanisms, will be investigated in greater detail in later chapters.

Unfortunately, there exist technical difficulties in investigating many of the hypotheses. Even if one is successful at isolating appropriate cell lines that exhibit the dormant behavior, typical experimental models involve mouse xenografts (implanting human-derived cancerous tissue into mice), which may or may not be applicable for these questions because of interspecies compatibilities.

Broadly speaking, mechanisms that can lead to tumor dormancy can be subdivided into *extrinsic* and *intrinsic* mechanisms.

Extrinsic mechanisms imply restrictions on growth, coming from the environment. For instance, changes in the tumor carrying capacity, whether due to variations in nutrient availability, acquisition of the ability to recruit blood vessels, and/or developing mechanisms to avoid immune surveillance, would allow the tumor to "escape" dormancy. Intrinsic mechanisms imply restrictions on growth that come from the structure of the population itself. For instance, solitary tumor dormancy, or quiescence, is characterized by lack of proliferation due to long periods of cell cycle arrest. While both mechanisms can lead to the phenomenon of latent tumor growth, it has been suggested based on a computer simulation study conducted by Wells and colleagues in 2013 that quiescence is more likely to be the mechanism responsible for metastatic dormancy.

In this chapter, we will focus on looking at the tumor dynamics before it reaches the carrying capacity allowed by its microenvironment. Specifically, we investigate several models that show that behavior consistent with escape from dormancy can arise as a result of natural growth dynamics of heterogeneous populations. We will look at heterogeneous populations that grow according to Malthusian, logistic, and Allee growth laws. We will demonstrate that dynamical behavior consistent with escape from tumor dormancy after a long period of latency can be generated due solely to the very heterogeneous nature of the populations described by these three models. We will also look at changes in population composition as these populations grow and develop and demonstrate what changes occur in the time period preceding escape from dormancy for populations that follow these growth laws.

MODELING PARAMETRICALLY HETEROGENEOUS POPULATIONS: PARAMETER DISTRIBUTION TECHNIQUE

To describe the dynamics of heterogeneous populations and how they evolve over time, we will use the parameter distribution technique, also known as the Reduction Theorem or the hidden keystone variable (HKV) method [20–22], which was developed by Georgy Karev at the NCBI, NIH. (The term "keystone" was chosen in analogue with keystone species in ecology, whose presence has a disproportionately large effect on the ecosystem relative in its abundance).

In brief, the technique involves subdividing the population into subpopulations, each of which is characterized by its own value of some intrinsic heritable parameter, such as growth rate, death rate, etc. These subpopulations are assumed to fall within some initial distribution. We can then introduce an auxiliary "keystone" variable that is chosen in such a way as to be tied to the moment generating function of the initial distribution of the subgroups within the population. A moment generating function allows us to track changes in expected value and variance of the chosen distributed parameter over time, thus allowing us to track the evolution of the population with respect to this parameter.

When considering the problems at hand, it is important to remember that the time horizon that we are looking at can be lengthy. For instance, based on estimates of when some people start smoking and when they may start showing signs of lung cancer, it has been estimated that tumors can take decades to grow from the initial DNA damage. Even survivors of Hiroshima and Nagasaki started showing increased incidences of cancers only in the 1950s [23–26]. Therefore, in our simulations, we will seek to reproduce dynamics consistent with escape from dormancy over an approximately 20-year period (240 months). Since in this chapter we are focusing on modeling tumor dynamics in the time preceding it reaching its carrying capacity, we will look at a simulated microscopic tumor reaching a size of approximately 1 mm^3, which is an approximate limit imposed on nonangiogenic tumor size by oxygen and nutrient constraints [18]. There are of course tumor- and individual-specific variations, but the goal of these simulations is to demonstrate that qualitative behavior of tumor dormancy and escape from it can be reproduced with these models.

PARAMETERS AND DISTRIBUTIONS

Choice of Initial Distribution

Whenever possible, the choice of initial distribution should be driven by data. Unfortunately, data are often not available, such as when we want to investigate the composition of a population of cancer cells before the tumor becomes clinically apparent. In this case, according to the principle of maximum entropy (MaxEnt), if the mean value of the random variable is the only quantity that can be estimated from observations or other data, then the most likely distribution of the variable is exponential with the estimated mean [27]. Since no value of a biological characteristic, such as a growth rate, can be infinite, we should choose the truncated exponential distribution as the initial distribution. So, for the numerical investigations in this chapter, the initial distribution is by default chosen to be truncated exponential on a finite interval (here we will mostly use interval [0,1]), unless indicated otherwise.

Parametrically Heterogeneous Malthusian Growth

Now let us work through the first example of a heterogeneous population that grows exponentially (also known as Malthusian growth, named after Thomas Robert Malthus, who described potential resource depletion problems that an exponentially growing population can encounter in his 1798 essay on population growth).

Consider a population of cells $x_c(t)$, where each individual cell is characterized by some intrinsic growth rate c. A set of cells that are characterized by the same value of c is referred to here as c-clone. The total size of the population is given by $N(t) = \sum_{\mathbb{A}} x_c(t)$ if c takes on discrete values and $N(t) = \int_{\mathbb{A}} x_c(t)dc$ if c is continuous. $\mathbb{A}$ denotes the range of possible values of parameter c.

If this population grows exponentially, then the rate of change of $x_c(t)$ is given by

$$x_c(t)' = cx_c(t). \tag{2.1}$$

Solution to this equation is $x_c(t) = e^{ct}x_c(0)$. The distribution of cell clones is by definition $P_c(t) = \frac{x_c(t)}{N(t)}$. Now let us plug in the aforementioned definition of $N(t)$, which results in the following equation:

$$N(t) = \int_{\mathbb{A}} x_c(t)dc = \int_{\mathbb{A}} x_c(0)e^{ct}dc = N(0)\int_{\mathbb{A}} e^{ct}\ P_c(0)\ dc = N(0)M_0[t], \tag{2.2}$$

where $M_0[t] = \int_A e^{ct}P_c(0)dc$ is by definition the moment generating function of the initial distribution of c-clones in the population.

It can be shown that the expected value of parameter c is given by

$$E^t[c] = \frac{dM_0[t]}{dt}\Big/ M_0[t] \tag{2.3}$$

and that variance of parameter c is given by

$$Var^t[c] = \frac{M_0[t]''}{M_0[t]} - \left(\frac{M_0[t]'}{M_0[t]}\right)^2. \tag{2.4}$$

The equation for the rate of change of the total population size thus becomes

$$\frac{dN}{dt} = E^t[c]\ N(t), \tag{2.5}$$

with $E^t[c]$ defined above.

Moment generating functions for many initial distributions can easily be found in probability theory or statistics textbooks. For truncated exponential distribution on the interval [0,1], the moment generating function is with parameter of exponential distribution μ and is given by

$$M_0[t] = \left(\frac{\mu}{e^{\mu} - 1}\right)\left(\frac{e^{\mu} - e^{t}}{\mu - t}\right), \tag{2.6}$$

the expected value of parameter c is

$$E^t[c] = \frac{e^t}{e^t - e^{\mu}} + \frac{1}{\mu - t} \tag{2.7}$$

and its variance is

$$Var^t[c] = \frac{1}{(\mu - t)^2} - \frac{e^{\mu + t}}{(e^{\mu} - e^t)^2}. \tag{2.8}$$

As mentioned above, we are looking to reproduce growth dynamics that begin with a period of very slow growth for a period of approximately 120–240 months (10–20 years), after which population size rapidly increases, reaching a volume of $\approx 1\ \text{mm}^3$.

In Fig. 2.1, we plot the changes over time of population size $N(t)$, the expected value (mean) of Malthusian parameter $E^t[c]$, and its variance $Var^t[c]$. As you can see, up until time $t \approx 180$, there is a period of very slow growth—the expected value $E^t[c]$, variance $Var^t[c]$, and the population size $N(t)$ increase very slowly. We will refer to this dynamics as *dormancy phase*. The dormancy phase is followed by a transitional *escape phase*, where population size increases very rapidly. In this example, where the parameter of truncated exponential distribution is $\mu = 200$, the escape phase occurs at $t \approx 180$ months = 15 years. During the escape phase, we can observe a characteristic bell-shaped graph of $Var^t[c]$, where variance increases together with population size $N(t)$ and the expected value $E^t[c]$, and then decreases again as the clones with the highest fitness (in this case the largest value of parameter c) become selected.

Now let us investigate what influences time to the escape phase within the framework of this model. It is reasonable to assume that if there are more rapidly growing clones in the population in the beginning (for instance, due to a greater mutational assault by carcinogens or weaker DNA repair mechanisms), the population will reach the escape phase sooner. In the model, variations in the initial composition of the

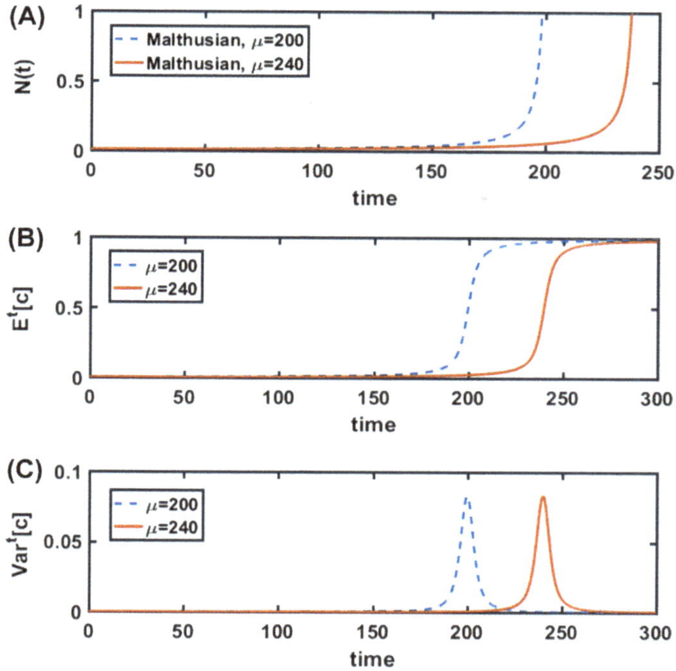

Figure 2.1 Parametrically heterogeneous Malthusian growth model with respect to growth rate parameter c. Initial distribution is taken to be truncated exponential, with distribution parameter $\mu = 200$ and $\mu = 240$. Initial population size is $N(0) = 0.01$. (A) Population of cells N(t), which remains at a negligible size until $t \approx 180$ for $\mu = 200$ and $t \approx 230$ for $\mu = 240$. (B) The expected value $E^t[c]$ of Malthusian parameter c increases rapidly concurrent with the escape phase of tumor growth. (C) Variance $Var^t[c]$ also increases dramatically during the escape phase but then decreases as the clone with the largest possible value of c becomes selected (in this case it is $E^t[c] \rightarrow 1$, since the exponential distribution was taken to be truncated exponential on the interval $c \in [0,1]$). One can clearly see that increase in the value of parameter of the initial distribution delays time to the escape phase, highlighting the importance of initial composition of the population in determining time to escape from dormancy.

population are taken into account with the value of parameter of the initial distribution.

This effect is also demonstrated in Fig. 2.1: larger value of μ corresponds to longer time to escape phase. The same effect is observed for population variance and the expected value: larger value of the parameter of the initial distribution corresponds to later spikes in the expected value of the distributed parameter and later appearance of the "bell shape" of population variance.

This simple model is already capable of reproducing dynamical behavior that is consistent with tumor dormancy and subsequent escape from it. However, this model allows for unrestrained growth, without consideration for spatial or nutrient limitations, which invariably exist in a tissue. Therefore, while this model can reproduce the dynamics in the initial stages, it becomes much less informative over time. This renders a logistic growth model more applicable.

Parametrically Heterogeneous Logistic Growth

Here, we will apply the parameter distribution technique to the logistic model. There exist some significant differences in application of the HKV method, compared to the case of the Malthusian model, namely, a necessity to introduce an auxiliary or "keystone" equation.

Once again, consider a population of cell clones $x_c(t)$, where each individual cell is characterized by an intrinsic growth rate c. The equation that describes growth of this population is given by

$$x_c(t)' = cx_c(t)\left(1 - \frac{N(t)}{K}\right), \tag{2.9}$$

where $N(t)$ is the total population size and K is the carrying capacity.

Let us introduce an auxiliary variable $q(t)$, such that

$$\frac{dq(t)}{dt} = 1 - \frac{N(t)}{K}. \tag{2.10}$$

Then

$$\frac{dx_c(t)}{x_c(t)} = cq(t)' \tag{2.11}$$

and therefore

$$x_c(t) = x_c(0)\, e^{cq(t)}. \tag{2.12}$$

We can now calculate total population size to be

$$\begin{aligned} N(t) &= \int_{\mathbb{A}} x_c(t)dc = \int_{\mathbb{A}} x_c(0)e^{cq(t)}\, dc = N(0)\int_{\mathbb{A}} P_c(0)e^{cq(t)}\, dc \\ &= N(0)\, M_0[q(t)], \end{aligned} \tag{2.13}$$

which follows from the definition of moment generating function $M_0[q(t)]$ for the initial distribution $P_c(0)$. Since in this case, the moment generating function of the truncated exponential distribution on the interval [0,1] is $M_0\left[q\left(t\right)\right]=\left(\frac{\mu}{e^{\mu}-1}\right)\left(\frac{e^{\mu}-e^{q(t)}}{\mu-q(t)}\right)$, the final equation for $q(t)'$ becomes

$$\frac{dq(t)}{dt}=1-\frac{N(0)}{K}M_0\left[q(t)\right]=1-\left(\frac{N(0)}{K}\right)\left(\frac{\mu}{e^{\mu}-1}\right)\left(\frac{e^{\mu}-e^{q(t)}}{\mu-q(t)}\right), \quad (2.14)$$

which fully closes the system.

The distribution of clones over time is by definition given by

$$P_c(t)=\frac{x_c(t)}{N(t)}=\frac{x_c(0)e^{cq(t)}}{N(0)M_0[q(t)]}. \quad (2.15)$$

Let us now plot the changes over time of population size, expected value of c, and its variance in Fig. 2.2. As one can see, the qualitative

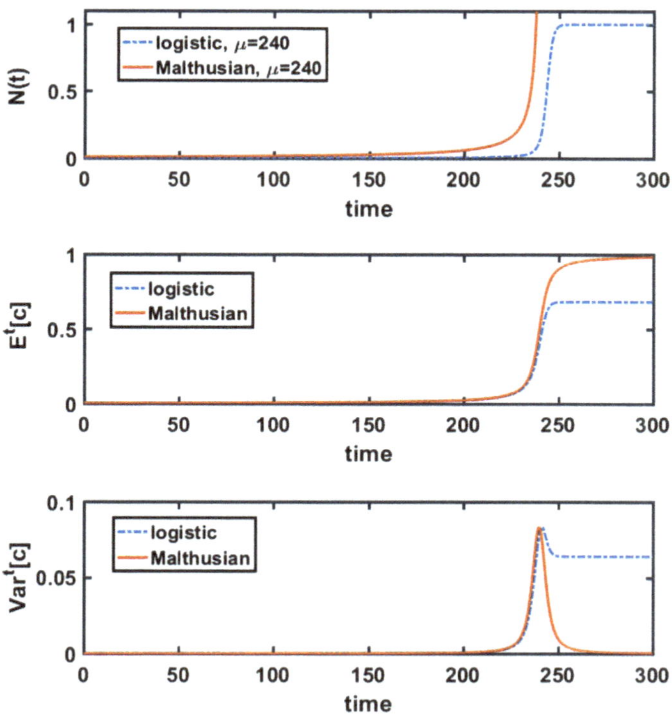

Figure 2.2 Comparison of distributed Malthusian and logistic growth models, with $N(0)=0.01$, and with truncated exponential initial distribution on the interval $c\in[0;1]$ with parameter of the distribution being $\mu=240$ for both models. For the logistic model, $K=1$. As one can see, dynamics up to the escape phase are identical but at the steady state, the logistic population maintains heterogeneity and consequently lower final $E^t[c]$.

behavior in the time preceding escape from dormancy is similar to that of the previous model; however, in this case, the population reaches its carrying capacity after the escape phase. (In this example, the carrying capacity if fixed; in later chapters, we will explore cases, when it can change as a result of tumor dynamics.)

The expected value $E^t[c]$ changes over time in a way that is very similar to the previous case, although notably here it does not reach the maximum value, allowed by the interval of the truncated exponential distribution, and remains below 1.

The dynamics of the variance $Var^t[c]$ is, however, qualitatively different from the previous case. It also increases dramatically in the moments preceding escape from dormancy. However, it remains at a nonzero value even after the population has reached its carrying capacity. Unlike the Malthusian model, in the parametrically heterogeneous logistic growth model, the population maintains heterogeneity at equilibrium and does not select for a single clone. Here we therefore do not observe selection of the fittest but instead *survival of everybody*.

This occurs because as $t \to \infty$, $q(t) \to q^*$, where q^* is some constant. Therefore, as $t \to \infty$, the population $N(t)$ grows like a Malthusian model but without tending to infinity, remaining at a finite moment in time, which allows maintaining population heterogeneity. One could conceivably interpret $q(t)$ as the "internal time" of the population and argue that if a tumor is growing according to a distributed logistic model, it lives according to its own "internal clock" that may be different from that of the surrounding tissues.

Parametrically Heterogeneous Allee Growth

Logistic model implies that the presence of any number of cells in the tumor population, however small, will give rise to a tumor. However, it has been reported that in a three-dimensional culture, depending on the environment and the surrounding signaling, at low concentrations malignant cells can sometimes revert to nonmalignant phenotype [28–30], suggesting the existence of some threshold of population viability. This can be accounted for by modifying the logistic growth model in the following way:

$$x_i' = cx_i(l - N)(N - m), \quad i = c, l, m \tag{2.16}$$

where $N(t)$ is the population size, l is the carrying capacity of the population, and m is the metastable point that divides areas of attraction of the two equilibria 0 and m. This type of model describes what is called the Allee effect, where the population tends to zero if its size is below viability threshold m and behaves like a logistic model if its size is above m.

Here, we will consider Allee-type models, heterogeneous with respect to each of the three parameters. We will first do the transformation for all three models and then compare their behavior.

Distribution of Growth Parameter c

Similarly to the previous cases, consider a population of clones $x_c(t)$ that are characterized by an intrinsic heritable value of parameter c. Introduce an auxiliary "keystone" variable $q_1(t)$ such that

$$q_1(t)' = (l - N(t))(N(t) - m). \tag{2.17}$$

Then

$$x_c(t) = x_c(0)\ e^{cq_1(t)}, \tag{2.18}$$

$$N(t) = N(0)M_0[q_1(t)], \tag{2.19}$$

$$P_t[c] = P_0[c]\frac{e^{cq_2(t)}}{M_0[q_2(t)]}. \tag{2.20}$$

The expected value $E^t[c]$ and variance $Var^t[c]$ of the population are defined through the moment generating function of the initial distribution according to formulas (2.3) and (2.4), respectively.

Distribution of Carrying Capacity l

Now consider a population of cells x_l that differs in the value of the carrying capacity l. Rewrite Eq. (2.16) as

$$\begin{aligned}\frac{dx_l(t)}{dt} &= cx_l(t)(l - N(t))(N(t) - m)\\ &= cx_l(t)(l(N(t) - m) - N(t)(N(t) - m))dt\end{aligned} \tag{2.21}$$

Now define the following keystone variables:

$$\begin{aligned}\frac{dp}{dt} &= c(N(t) - m),\\ \frac{dq}{dt} &= cN(N(t) - m).\end{aligned} \tag{2.22}$$

Then $\frac{dx_l(t)}{dt} = x_l(t)\left(l\frac{dp}{dt} - \frac{dq}{dt}\right)$ and consequently

$$x_l(t) = x_l(0)\ e^{lp(t)-q(t)}. \tag{2.23}$$

Total population size then is given by

$$N(t) = \int_L x_l(t)dl = \int_L x_l(0)\ e^{lp(t)}\ e^{-q(t)}dl = N_0\ e^{-q(t)}\int_L P_0(l)\ e^{lp(t)}\,dl$$
$$= N_0 e^{-q(t)} M_0[p(t)]. \tag{2.24}$$

The distribution of clones is

$$P_l[t] = \frac{x_l(t)}{N(t)} = \frac{x_l(0)\ e^{lp(t)-q(t)}}{N_0\ e^{-q(t)}\ M_0[p(t)]} = P_l[0]\frac{e^{lp(t)}}{M_0[p(t)]}. \tag{2.25}$$

The expected value of l is $E^t[l] = \frac{M_0[p(t)]'}{M_0[p(t)]}$.

The characteristics of the full system can thus be given by Eq. (2.17), with $N(t)$ defined above by Eq. (2.18).

Distribution of Parameter m

Finally, consider the case, where each clone is characterized by an individual value of parameter m. Here, the keystone equations are

$$\begin{aligned} q(t)' &= Nc(l - N),\\ p'(t) &= -c(l - N). \end{aligned} \tag{2.26}$$

Consequently, $\frac{x_m(t)'}{x_m(t)} = q(t)' + mp(t)'$ and $x_m(t) = x_m(0)e^{q(t)+mp(t)}$.

The total population size is given by

$$\begin{aligned} N(t) &= \int_{\mathbb{M}} x_m(0)\ e^{q(t)}\ e^{mp(t)}dm = \int_{\mathbb{M}} P_0(m)\ N(0)\ e^{q(t)}\ e^{mp(t)}dm\\ &= N(0)\ e^{q(t)}\int_{\mathbb{M}} P_m(0)\ e^{mp(t)}dm = N(0)\ e^{q(t)}M_0[p(t)]. \end{aligned} \tag{2.27}$$

Here is how all three models compare to each other:

Original system $\quad x_i' = cx_i(l-N)(N-m),$

	***c*-distributed**	***l*-distributed**	***m*-distributed**
Keystone equations	$q_1(t)' = (l - N(t)) \times (N(t) - m)$	$p'(t) = c(N(t) - m),$ $q'(t) = cN(t)(N(t) - m).$	$q(t)' = N(t)\ c(l - N(t)),$ $p'(t) = -c(l - N(t)).$
Total population size $N(t)$	$N(t) = N(0) \times M_0[q_1(t)]$	$N(t) = N_0\ e^{-q(t)}M_0[p(t)].$	$N(t) = N(0)\ e^{q(t)} \times M_0[p(t)].$

As one can see, the same original system, depending on which parameter is distributed, can have a very different final form, which will result in different dynamics.

All three models provide dynamical behaviors that are consistent with escape from dormancy (Fig. 2.3). However, escape phase predicted by c-distributed Allee model occurs in the most gradual way out of all of the examples (Fig. 2.3A) and is accompanied by slight increase in the expected value of c (Fig. 2.3B) and a steady increase in $Var^t[c]$ (Fig. 2.3C) at equilibrium.

A population that grows according to l-distributed Allee model exhibits sharper increase in population size during the escape phase; its final population composition is the most homogeneous, with selection toward the largest value of l.

Finally, a population that grows according to the m-distributed Allee model exhibits a more unusual dynamics, with population size $N(t)$ dropping to near zero and remaining at that level for many months and even years before eventually rapidly increasing in size (Fig. 2.3A). Like in every case described, the escape phase in all of the populations is accompanied by increase in both expected value and variance of the distributed parameter (Fig. 2.3B). But in this case, the population becomes less heterogeneous (see change in variance) over time (Fig. 2.3C); however, it loses heterogeneity much more gradually, compared to population with distributed carrying capacity l.

PRIMARY DORMANCY, METASTATIC DORMANCY, BOTH OR NEITHER?

We have shown here that several models of parametrically heterogeneous populations can reproduce behaviors that are consistent with what we expect to see in tumor dormancy and escape from it. However, which model is the right one or at least the one closest to "reality"?

Just looking that the population growth dynamics does not provide the answer—we need to look at the other characteristics, namely the mean of the distributed parameters and their variance.

One thing that is known about most tumors is that they are heterogeneous and mostly remain heterogeneous over time, at least to some degree; therefore, the models that predict tumor homogeneity are unlikely to be applicable. These include Malthusian growth model and the l-distributed Allee-type model.

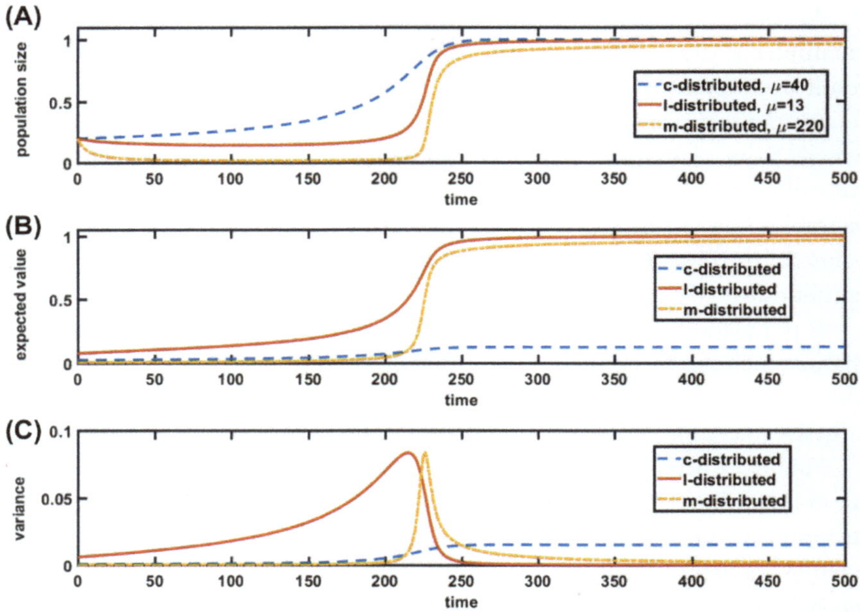

Figure 2.3 Comparison of the three parametrically heterogeneous Allee growth models. The initial distribution is truncated exponential on the interval [0,1] for all three models. Parameters are $c = 1, l = 1, m = 0.1$ for each respective model; the initial population size is $N(0) = 0.2$. (A) The total population size $N(t)$ for all three models, with c-distributed Allee model increasing most gradually, l-distributed Allee model remaining at a steady size before increasing, and m-distributed Allee model predicting a decrease in population size to near zero until the escape phase. (B) Expected value of each of the parameters for their respective models and (C) variance of each of the parameters for their respective models. In every case, escape phase is accompanied by increase in both the expected value and variance. The c-distributed model maintains heterogeneity at a steady state; l-distributed model rapidly becomes more homogeneous at the steady state; m-distributed model gradually loses heterogeneity over time after the escape phase.

What about the logistic model and the *m*-distributed and *c*-distributed Allee-type models?

The heterogeneous logistic model predicts a rapid escape phase and a nonzero population variance at the equilibrium state. If this model were to describe the dynamics of a dormant tumor, it is possible that it could describe either a primary or a secondary tumor that is subject to environmental constraints of the tissue.

In the *c*-distributed Allee model, the escape phase is very gradual, suggesting that it might be more applicable to describe dynamics of slowly growing tumors, such as some cases of prostate cancer. Finally, the *m*-distributed Allee model predicts that a population size can dip to near zero for months or even years and then regrow again, suggesting that it would be most appropriate to describe metastatic dormancy. Furthermore, the decrease in population heterogeneity that occurs at the equilibrium state is consistent with more gradual selection for clones with highest fitness. In this case, the tumor eventually becomes dominated by a single clone, a scenario that can occur in metastatic tumors. These considerations are summarized in Table 2.2.

Interestingly, a number of experimental models have suggested that metastatic spread can occur at very early stages of primary tumor development [31]. In their 2014 paper, Newman and Cisneros [32] investigated the question of whether, from a theoretical point of view, dissemination using few very specialized cells (special forces) or seeding a large number of nonspecialized cells (infantry) would be more effective for metastatic spread. Their analysis showed that in fact both strategies are equally likely (or unlikely) to seed disseminated tumors. Therefore, it is possible that a tumor that has disseminated early might be growing according to the same laws in an unprimed environment, much like a primary dormant tumor, and escape dormancy independently of the primary tumor or any other environmental factors, simply growing according to its own clock.

GETTING BACK TO THE QUESTION

So far we have looked at the phenomenon of tumor dormancy, and particularly preclinical dormancy, i.e., when microscopic tumors persist but do not become clinically apparent, "cancer without disease." We have shown that perhaps different tumors in different tissues may be described by different growth laws, which may be reflecting both the nature of the underlying tumor and constraints imposed by the environment (i.e., variability in growth rates, death rates, sensitivities to various components of the microenvironment, such as hormone availability, metabolic heterogeneity, etc.). These tumors reach an "escape"

Table 2.2 Summary of results and possible interpretation of the applicability of various parametrically heterogeneous models to describing the dynamics underlying tumor dormancy

Model	Distributed parameter	Escape phase	Expected value of distributed parameter	Variance of distributed parameter after escape	Interpretation/ applicability
Malthusian	c	Rapid	Increases rapidly during escape phase to maximum possible value	Increases rapidly during escape, then returns to zero	Likely not applicable
Logistic	c	Rapid	Increases rapidly during escape phase to submaximum value (inversely proportional to variance)	Increases rapidly during escape, then remains at a nonzero value	Both primary and metastatic tumor dormancy
Allee $x_c{}' = cx_c(l - N)(N - m)$,	c	Gradual	Increases rapidly during escape phase to submaximum value (inversely proportional to variance)	Increases rapidly during escape, then remains at a nonzero value	Slowly growing tumor
$x_l{}' = cx_l(l - N)(N - m)$,	l	Less gradual	Increases rapidly during escape phase to maximum possible value	Increases rapidly during escape, then returns to zero	Likely not applicable
$x_m{}' = cx_m(l - N)(N - m)$,	m	Rapid	Increases rapidly during escape phase to submaximum value (inversely proportionate to variance)	Increases rapidly during escape, then gradually decreases	Metastatic tumor dormancy

This table was adapted from Kareva, I. Primary and metastatic tumor dormancy as a result of population heterogeneity. Biology Direct 2016;11(1):37.

phase according to their own time schedule, uninfluenced by the environment.

It is possible that just by chance, this is why secondary tumors may start growing after removal of primary tumors—it is just that their time to escape phase has arrived. While this may happen in a small subset of cases, it is not very likely to be responsible for the majority, since nothing about primary tumor resection would trigger the escape phase if the tumor is just growing according to its own time clock. Therefore, it is not likely that the "tumor time bomb" theory explains why removal of primary tumors may trigger accelerated growth of secondary tumors.

In the next chapter we will draw our attention to tumor–immune interactions. Perhaps that is where we can find clues to answer our question.

REFERENCES

[1] Nielsen M, Thomsen J, Primdahl S, Dyreborg U, Andersen J. Breast cancer and atypia among young and middle-aged women: a study of 110 medicolegal autopsies. British Journal of Cancer. Nature Publishing Group 1987;56(6):814.
[2] Sakr W, Grignon D, Crissman J, Heilbrun L, Cassin B, Pontes J, et al. High grade prostatic intraepithelial neoplasia (HGPIN) and prostatic adenocarcinoma between the ages of 20–69: an autopsy study of 249 cases. In Vivo (Athens, Greece) 1993;8(3):439–43.
[3] Fife BT, Bluestone JA. Control of peripheral T-cell tolerance and autoimmunity via the CTLA-4 and PD-1 pathways. Immunological Reviews. Wiley Online Library 2008;224(1):166–82.
[4] Naumov GN, MacDonald IC, Chambers AF, Groom AC. Solitary cancer cells as a possible source of tumour dormancy? Seminars in Cancer Biology. Elsevier 2001:271–6.
[5] Folkman J, Kalluri R. Cancer without disease. Nature. Nature Publishing Group 2004;427(6977):787.
[6] Klein CA, Blankenstein TJ, Schmidt-Kittler O, Petronio M, Polzer B, Stoecklein NH, et al. Genetic heterogeneity of single disseminated tumour cells in minimal residual cancer. The Lancet. Elsevier 2002;360(9334):683–9.
[7] Klein CA. Framework models of tumor dormancy from patient-derived observations. Current Opinion in Genetics & Development. Elsevier 2011;21(1):42–9.
[8] Weckermann D, Polzer B, Ragg T, Blana A, Schlimok G, Arnholdt H, et al. Perioperative activation of disseminated tumor cells in bone marrow of patients with prostate cancer. Journal of Clinical Oncology. American Society of Clinical Oncology 2009;27(10):1549–56.
[9] Luzzi KJ, MacDonald IC, Schmidt EE, Kerkvliet N, Morris VL, Chambers AF, et al. Multistep nature of metastatic inefficiency: dormancy of solitary cells after successful extravasation and limited survival of early micrometastases. The American Journal of Pathology. Elsevier 1998;153(3):865–73.

[10] Strauss DC, Thomas JM. Transmission of donor melanoma by organ transplantation. The Lancet Oncology. Elsevier 2010;11(8):790–6.
[11] Buell JF, Trofe J, Hanaway MJ, Lo A, Rosengard B, Rilo H, et al. Transmission of donor cancer into cardiothoracic transplant recipients. Surgery. Elsevier 2001;130(4):660–8.
[12] Woodle E, Gupta M, Buell J, Neff G, Gross T, First M, et al. Prostate cancer prior to solid organ transplantation: the Israel Penn International Transplant Tumor Registry experience. Transplantation Proceedings. Elsevier 2005:958–9.
[13] Aguirre-Ghiso JA. Models, mechanisms and clinical evidence for cancer dormancy. Nature Reviews Cancer. Nature Publishing Group 2007;7(11):834–46.
[14] Almog N. Molecular mechanisms underlying tumor dormancy. Cancer Letters. Elsevier 2010;294(2):139–46.
[15] Wilkie KP, Hahnfeldt P. Tumor–immune dynamics regulated in the microenvironment inform the transient nature of immune-induced tumor dormancy. Cancer Research. AACR 2013;73(12):3534–44.
[16] Teng MW, Swann JB, Koebel CM, Schreiber RD, Smyth MJ. Immune-mediated dormancy: an equilibrium with cancer. Journal of Leukocyte Biology. Society Leukocyte Biology 2008;84(4):988–93.
[17] DeGregori J. Evolved tumor suppression: why are we so good at not getting cancer? Cancer Research. AACR 2011;71(11):3739–44.
[18] Naumov GN, Akslen LA, Folkman J. Role of angiogenesis in human tumor dormancy: animal models of the angiogenic switch. Cell Cycle. Taylor & Francis 2006;5(16):1779–87.
[19] Naumov GN, Folkman J, Straume O. Tumor dormancy due to failure of angiogenesis: role of the microenvironment. Clinical & Experimental Metastasis. Springer 2009;26(1):51–60.
[20] Karev GP. Dynamics of inhomogeneous populations and global demography models. Journal of Biological Systems. World Scientific 2005;13(01):83–104.
[21] Karev G, Kareva I. Replicator equations and models of biological populations and communities. Mathematical Modelling of Natural Phenomena 2014;9(3):68–95.
[22] Karev GP. On mathematical theory of selection: continuous time population dynamics. Journal of Mathematical Biology. Springer 2010;60(1):107–29.
[23] Pierce DA, Preston DL. Radiation-related cancer risks at low doses among atomic bomb survivors. Radiation Research 2000;154(2):178–86.
[24] Thompson DE, Mabuchi K, Ron E, Soda M, Tokunaga M, Ochikubo S, et al. Cancer incidence in atomic bomb survivors. Part II: solid tumors, 1958–1987. Radiation Research. Radiation Research Society 1994;137(2s):S17–67.
[25] Preston DL, Kusumi S, Tomonaga M, Izumi S, Ron E, Kuramoto A, et al. Cancer incidence in atomic bomb survivors. Part III: leukemia, lymphoma and multiple myeloma, 1950–1987. Radiation Research. Radiation Research Society 1994;137(2s):S68–97.
[26] Land CE, Tokunaga M, Koyama K, Soda M, Preston DL, Nishimori I, et al. Incidence of female breast cancer among atomic bomb survivors, Hiroshima and Nagasaki, 1950–1990. Radiation Research 2003;160(6):707–17.
[27] Kapur JN. Maximum-entropy models in science and engineering. John Wiley & Sons; 1989.

[28] Bissell MJ, Radisky DC, Rizki A, Weaver VM, Petersen OW. The organizing principle: microenvironmental influences in the normal and malignant breast. Differentiation 2002;70(9–10):537–46.
[29] Lee GY, Kenny PA, Lee EH, Bissell MJ. Three-dimensional culture models of normal and malignant breast epithelial cells. Nature Methods 2007;4(4):359–65.
[30] Weaver VM, Petersen OW, Wang F, Larabell CA, Briand P, Damsky C, Bissell MJ. Reversion of the malignant phenotype of human breast cells in three-dimensional culture and in vivo by integrin blocking antibodies. The Journal of Cell Biology 1997;137(1):231–45.
[31] Hüsemann Y, Geigl JB, Schubert F, Musiani P, Meyer M, Burghart E, et al. Systemic spread is an early step in breast cancer. Cancer Cell. Elsevier 2008;13(1):58–68.
[32] Cisneros LH, Newman TJ. Quantifying metastatic inefficiency: rare genotypes versus rare dynamics. Physical Biology. IOP Publishing 2014;(4):11, 046003.

CHAPTER 3

Cancer Immunoediting: A Process Driven by Metabolic Competition as a Predator–Prey-Shared Resource Type Model

Contents

BACKGROUND

One of the mechanisms halting tumor growth is management by the immune system. As the cells undergo malignant transformation as a result of carcinogenic mutations (changing shape, acquiring large, variably shaped nuclei, abnormal membranes, disrupting tissue's regular organization), they may become recognized by the body's innate immune cells, particularly natural killer cells. The debris of cancer cells can aid in activating cells of the adaptive immune response, such as CD8+T cells, which then become the dominant force in charge of controlling growth of a malignant tumor. The CD8+T cells bind to receptors on cancer cells' surface and release chemicals

Understanding Cancer From a Systems Biology Point of View
ISBN 978-0-12-813673-7
http://dx.doi.org/10.1016/B978-0-12-813673-7.00003-3

such as perforin that damage cancer cell membrane. This creates an osmotic imbalance inside the cell that forces it to either swell or shrink. After (hopefully) eliminating cancer cells, a fraction of the cells of the adaptive immune response remains in the body for extended periods of time. These memory T cells can then accelerate reactivation of the immune system, should the tumor reappear [1].

The immune system can be very efficient at cancer elimination, so why does it sometimes fail? One possibility is that cancer cells mutate too quickly for the immune system to catch up. Another possibility is that cancer cells look so much like the body's normal cells, expressing no abnormal antigens, that the immune cells simply do not recognize it as being foreign. However, there is another mechanism that can play a critical role in the way that cancer cells evade recognition by the immune system, and it has to do with metabolism.

It is a well-established fact that cancer cells can significantly upregulate glucose consumption to meet the high energetic demands for cell survival and proliferation. In fact, cancer cells use fast but less efficient anaerobic mode of glucose metabolism even in the areas of ample oxygen supply, a phenomenon that has become known as Warburg effect, which yields 2 molecules of ATP per glucose compared to approximately 30 ATPs that would have been obtained as a result of oxidative phosphorylation [2]. (Noticeably, in the literature, it is often cited that 36 molecules of ATP are produced as a result of aerobic metabolism; however, according to Berg et al. [3], this number is an overestimation and in fact between 26 and 30 ATPs are formed when glucose is oxidized completely.) A switch to glycolysis as a primary mode of glucose metabolism is often accompanied by up to 30-fold increase in glucose uptake, as well as overexpression of glucose transporters, such as GLUT-1 and SGLT-1, which is a natural adaptation for tumors that require larger intake of glucose to meet their metabolic needs [4].

What is often overlooked, however, is that actively proliferating cytotoxic lymphocytes also switch to purely glycolytic mode of glucose metabolism, which is accompanied by a similar upregulation of the expression of glucose transporters, such as GLUT-1. Inability to secure adequate nutrient supply during lymphocyte activation prevents cell growth and limits cell functionality, as cells deprived of adequate glucose supply cannot produce cytokines such as IFN-γ, which are required for effector function [5]. Therefore, it is not unreasonable to assume that in the tumor microenvironment there can exist competition for glucose between tumor and immune cells, and the outcome of this interaction would be determined by whichever cell type is the first to acquire access to glucose.

These considerations allow to conceive of the following theoretical setup: as the primary tumor increases, it may start outgrowing its oxygen supply. This forces it to use glycolysis as a primary mode of glucose metabolism, effectively selecting for cells with highly upregulated nutrient transporters. Eventually this may lead creation of a "glycolytic core" inside the tumor, while the cells on the outside of the tumor may still continue using more energy-efficient aerobic metabolism due to easier access to oxygen. Assuming sufficient tumor immunogenicity and functional immune response, actively proliferating immune cells that are attracted to the tumor site would eliminate the aerobic cells outside the tumor, exposing the glycolytic core. The newly exposed tumor cells, however, have been under selective pressure for highly upregulated nutrient transporters. This adaptation enables them to efficiently drain the shared glucose from the microenvironment, effectively outcompeting cytotoxic lymphocytes, which cannot perform their function without adequate nutritional support. As a result, the tumor cells can continue to grow without interference from nutritionally incapacitated immune cells. A schematic representation of the proposed scenario is given in Fig. 3.1.

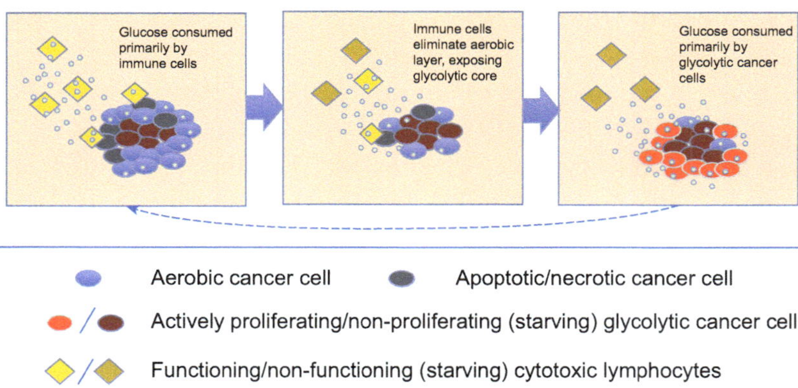

Figure 3.1 ***Proposed scenario of metabolism-driven tumor escape.*** In the tumor microenvironment, glucose is taken up by cytotoxic lymphocytes that use glycolysis as a primary mode of metabolism. "Peeling" off the cells on the tumor periphery and revealing its glycolytic core exposes glycolytic tumor cells with upregulated nutrients transporters that can now compete effectively with immune cells for resources. Lymphocytes cannot proliferate unless their nutrient demands are met, which can allow tumor cells to circumvent immune response, leading to tumor escape.

This scenario is an example of a hypothesis that can be tested using a mathematical model—if our assumptions are correct, what kind of behavior can we expect from such a system?

For this case, we will use a four-dimensional system of ordinary differential equations describing a predator–prey common resource relationship, where the cytotoxic lymphocytes are the predator, cancer cells are the prey, and glucose is the common resource. We hypothesize that access to glucose can influence tumor–immune interactions.

Notably, the goal here is not to provide a full analysis of the system, which can be found in Ref. [6]. Instead, the aim is to demonstrate how a mechanism-driven mathematical model is written and approached, and how it can be used to ask questions about the effects of therapeutic interventions.

MODEL DESCRIPTION

The population of tumor cells $T(t)$ is divided into two subpopulations of aerobic and glycolytic cells such that $T(t) = T_a(t) + T_g(t)$. The dynamics of the two subpopulations is described by identical equations that differ solely in parameter values. Tumor growth is assumed to be proportional to the amount of glucose that is available to the cells in the microenvironment, which is described by the term $r_j T_j(t) \frac{G(t)}{\xi_T + b_T G(t)}$, where $j = a$ represents aerobic cells and $j = g$ represents glycolytic cells; this functional form was taken from classical models of consumer–resource interactions, reviewed in Cantrell et al. [7]. Tumor cells can die either naturally at a rate $\mu_j T_j(t)$ or they can be killed by cytotoxic lymphocytes $I(t)$, which is accounted for by the term $e_j T_j(t) \frac{I(t)/T_j(t)}{s + I(t)/T_j(t)}$; this functional form is taken from models describing tumor–immune interactions developed by de Pillis et al. [8,9].

Glucose $G(t)$ is replenished from the bloodstream at a constant rate G_0 and is consumed by tumor cells according to functional form $d_j T_j(t) \frac{G(t)}{\xi_T + b_T G(t)}$ and by immune cells $d_I I(t) \frac{G(t)}{\xi_I + b_I G(t)}$. Since actively proliferating cytotoxic immune cells use glycolysis as a primary model of glucose metabolism, we assume that immune cells and glycolytic tumor cells deplete shared glucose supplies at approximately the same rate, i.e., $d_I \approx d_g$; the rate of glucose consumption by aerobic cells r_a however has been estimated to be up to 20–30 times lower. Glucose is also consumed by normal tissues, which in our model is accounted for by the term $\mu_G \frac{G(t)}{\xi_N + b_N G(t)}$. We assume here $b_N \ll 1$, which allows to approximate this term as $\mu_G G(t)$.

Finally, cytotoxic immune cells are recruited to the tumor site at a rate $i_0 T(t)$, which is a composite term representing tumor-induced recruitment of the activated immune cells to the tumor microenvironment. Depending on the question, this term can be "unpacked" to describe, for instance, the process of immune cell activation in response to activity of the innate immune system. Parameter i_0 can vary significantly depending on the degree of systemic immune stimulation or tumor-induced suppression; it can also be affected by various types of immunotherapy, a question that will be explored later in the chapter.

Immune cells are assumed to die at some natural rate $\mu_I I(t)$, which can also vary depending both on overall health of the patient and on various external stimuli. Finally, expansion of cytotoxic immune cells depends both on the number of tumor cells that have already been killed by the immune cells and on the amount of glucose available to the immune cells in the tumor microenvironment. This restriction is imposed to account for the fact that due to their high energy demands, even fully activated cytotoxic immune cells cannot perform their function if they do not have access to sufficient amounts of glucose [5].

The assumptions that were used in model formulation can be summarized as follows:

1. Tumor population consists of aerobic and glycolytic cells.
2. Glycolytic tumor cells not only grow faster but also deplete common resources faster (they have higher nutrient uptake rates).
3. Glucose is depleted at different rates by tumor and immune cells: aerobic cells deplete it slower, glycolytic tumor cells and immune cells deplete it faster.
4. Population of immune cells decreases in the absence of tumor, since cytotoxic lymphocytes need costimulatory signals from tumor cells to enable them to take up glucose; otherwise they die by starvation.
5. Expansion and functionality of immune cells is glucose dependent: they cannot grow and perform their function if they do not have access to sufficient amounts of glucose.

Taking all of these assumptions into account, we end up with the following system of ordinary differential equations:

$$
\begin{aligned}
\frac{dT_a}{dt} &= \underbrace{r_a T_a(t)\frac{G(t)}{\xi_T + b_T G(t)}}_{\text{glucose-dependent growth}} - \underbrace{e_a \frac{T_a(t)I(t)}{I + s(T_a(t) + T_g(t))}}_{\text{ratio-dependent death by CTLs}} - \underbrace{\mu_a T_a(t)}_{\text{natural death}} \\
\frac{dT_g}{dt} &= r_g T_g(t)\frac{G(t)}{\xi_T + b_T G(t)} - e_g \frac{T_g(t)I(t)}{I(t) + s(T_a(t) + T_g(t))} - \mu_g T_g(t) \\
\frac{dG}{dt} &= \underbrace{(G_0 - \mu_G G(t))}_{\text{natural G inflow/outflow}} - \underbrace{(d_a T_a(t) + d_g T_g(t))\frac{G(t)}{\xi_T + b_T G(t)}}_{\text{G consumed by the tumor}} - \underbrace{d_I I(t)\frac{G(t)}{\xi_I + b_I G(t)}}_{\text{G consumed by immune cells}} \\
\frac{dI}{dt} &= \Big(\underbrace{i_0(T_a(t) + T_g(t))}_{\text{systemic immune stimulation}} - \underbrace{\mu_I I(t)}_{\text{natural decay}} \Big) + \underbrace{r_I I(t)\left(\frac{G(t)}{\xi_I + b_I G(t)}\right)\left(\frac{I(t)(T_a(t) + T_g(t))}{I(t) + s(T_a(t) + T_g(t))}\right)}_{\substack{\text{cell expansion stimulated by debris from previously} \\ \text{killed tumor cells and modulated by glucose}}}
\end{aligned}
\tag{3.1}
$$

Description of the parameters and sample values are given in Table 3.1.

MODEL INVESTIGATION: USING THE MODEL TO ANSWER QUESTIONS

Brief Introduction to Dynamical Regimes

Firstly, we wish to identify the possible dynamical regimes within this model.

Generally, within dynamical systems there exist three main classes of dynamical regimes: stable equilibrium, unstable equilibrium, and oscillatory regimes (both stable and unstable). Equilibria are found by setting the equation to zero (when the rate of change is zero, the population is at equilibrium). Stable equilibrium implies that the system will always return to it after small disturbances. Unstable equilibrium implies that unless the system starts precisely at it, it will move away from it after small disturbances. Finally, oscillations can circumscribe the stable or unstable equilibria: if there is a stable equilibrium inside unstable oscillatory regime (called limit cycle), then, as long as the initial conditions are within the limit cycle, the system will return to the stable equilibrium. If the limit cycle is stable but the equilibrium point within it is unstable, then, as long as the

Table 3.1 Variables and parameters used in Eq. (3.1)

	Description	Estimated value	Units
$T(t)$	Population of tumor cells, composed of aerobic $T_a(t)$ and glycolytic $T_g(t)$ cells	$T(t) \geq 0$	vol.
$G(t)$	Glucose	$G(t) \geq 0$	moles
$I(t)$	Cytotoxic immune cells	$I(t) \geq 0$	vol.
r_a	Growth rate of aerobic cancer cells	4.31×10^{-1}	1/day
r_g	Growth rate of glycolytic cancer cells	$r_a \times k_1, k_1 \in [0.5, 5.0]$	1/day
μ_a	Natural death rate or aerobic cancer cells	0.01	1/day
μ_g	Natural death rate of glycolytic cancer cells	$\mu_g \times k_2, k_2 \in [1.0, 10.0]$	1/day
e_a	Fractional aerobic cell kill by immune cells	0.1	1/day
e_g	Fractional glycolytic cell kill by immune cells	$e_a - k_3, k_3 \in [-0.1, 0.1]$	1/day
s	Steepness coefficient of the tumor cell kill by immune cells	6.18×10^{-1}	n/a
b_T	Steepness coefficient for glucose-dependent growth of tumor cells	0.9	n/a
b_I	Steepness coefficient for glucose-dependent growth of immune cells	0.9	n/a
G_0	Rate of glucose inflow into tumor microenvironment from the bloodstream	$G_0 \in [0.01, 2.06]$	mol/day
μ_G	Rate of glucose outflow from tumor microenvironment from bloodstream	0.01	1/day
d_a	Rate of glucose consumption by aerobic tumor cells	$d_a \in [0.048, 0.5]$	mol/vol/day
d_g	Rate of glucose consumption by glycolytic tumor cells	$d_g \in [0.1268, 0.5262]$	mol/vol/day
d_I	Rate of glucose consumption by actively proliferating immune cells	$d_I \approx d_g$	mol/vol/day
i_0	Rate of tumor-stimulated inflow of immune cells	$i_0 \in [0.001, 0.100]$	vol/day
μ_I	Natural death rate of cytotoxic immune cells	2.0×10^{-2}	1/day
r_I	Rate at which immune cells are stimulated to be produced as a result of coming in contact with tumor debris from previously killed tumor cells	3.03×10^{-1}	1/vol/day
ξ_I, ξ_T	Saturation coefficient for glucose-dependent immune and tumor growth	1	moles

Owing to the form in which data were available, we chose to cast $G(t)$ in the units of moles as opposed to volume; the values of parameters that compare dynamics of aerobic as compared to glycolytic cells were taken based on theoretical considerations and citations.

initial conditions fall within the limit cycle, the system will stably oscillate indefinitely (we will see an example of it later in the chapter).

There are of course many more variations of these regimes, and the study of how systems change stability with respect to variations in parameter values is the subject of a fascinating field known as bifurcation theory. This overview however is sufficient to answer the questions posed in this chapter.

Aerobic and Glycolytic Tumor Cells Can Cooperate to Defeat the Immune System

Investigation of dynamical systems involves identifying parameters of interest and varying them one by one to find out how these variations can affect predicted system dynamics.

Here, we will focus on the effect of parameter e_j (effectiveness of tumor elimination by the immune system, $j = a, g$) on system dynamics since the ability of the immune system to recognize cancer is the key determinant of the effectiveness of tumor elimination by the immune system. We also look at the effects of variations of parameters i_0 (systemic immune stimulation) and μ_I (death rate of cytotoxic lymphocytes) because these parameters can, at least in theory, be manipulated in an experimental setting via various therapeutic interventions.

Unless otherwise indicated, parameters were held constant at $b_I = b_T = 0.9$, $e_a = e_g = 0.1$, $r_a = 0.1$, $r_g = 1.9r_a$, $s = 0.7$, $d_a = 0.1$, $d_g = d_I = 2d_a$, $\mu_a = 0.01$, $\mu_g = 3\mu_a$, $G_0 = 1$, $\mu_I = 0.01$, $r_I = 0.01$, ξ_T, ξ_I, $\xi_N = 1$, $T(0) = I(0) = C(0) = 1$, $T_a(0) = 0.1T$. Parameter values were chosen based on the values provided in the validated mathematical models proposed by de Pillis et al. [14,15]; in other cases, parameter values were estimated based on theoretical considerations. We chose to keep the units of volume unspecified to expand applicability of the model beyond the current framework, since the conceptual structure of the model as a predator–prey-shared resource type system might exist in other areas of ecology. As was mentioned in the introduction, mathematical modeling is very much about seeing relationships, which are transferrable between various fields.

In the first set of simulations, we varied parameter i_0 and were able to observe several of the dynamical regimes described above. As we increased i_0, we observed the dynamical shift from tumor persistence (Fig. 3.2A) to multiple cyclical regimes (Fig. 3.2B) to decrease in cycle number to tumor size stabilizing at a low equilibrium value (Fig. 3.2C), and then finally to the

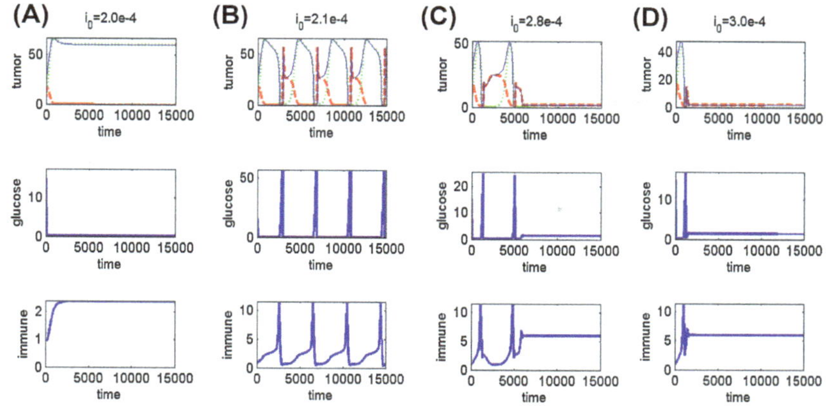

Figure 3.2 Different dynamical regimes that can be observed through progressively increasing parameter i_0, which represents systemic immune stimulation. Some possible regimes that can be observed range from (A) escape of a primarily aerobic tumor to (B) sustained cyclical behavior described in Fig. 3.1 to (C) several iterations of cyclical behavior before bringing down tumor size to (D) tumor control.

tumor reaching an equilibrium state with the immune system without any preceding cyclical dynamics (Fig. 3.2D).

Within this set of simulations the cyclical dynamics can recur several times before the immune system brings down the overall tumor size. Moreover, the composition of the tumor with respect to metabolic strategy can differ at different stages of the cycle! For instance, while in Fig. 3.2A the tumor is composed primarily of aerobic cells, in Fig. 3.2C and D, the tumor is composed of glycolytic cells.

The specifics of the dynamics that drive the oscillations can be observed in more detail in Fig. 3.3A, which is a magnification of the dynamics shown in Fig. 3.2B.

As we can see, predictably, increase in the number of immune cells corresponds to decrease in the number of aerobic tumor cells. As tumor size decreases, so does the population of immune cells, since constimulatory molecules from tumor debris are necessary for clonal expansion of immune cells (this is accounted for by the last term in the equation for dI/dt). However, as the population of immune cells decreases and stops utilizing glucose, its levels rise and it can now be then taken up by expanding population of glycolytic tumor cells. This causes a sharp drop in glucose levels in the tumor microenvironment, preventing immune cells from proliferating, in this case not from lack of costimulation but from lack of resources necessary for clonal expansion.

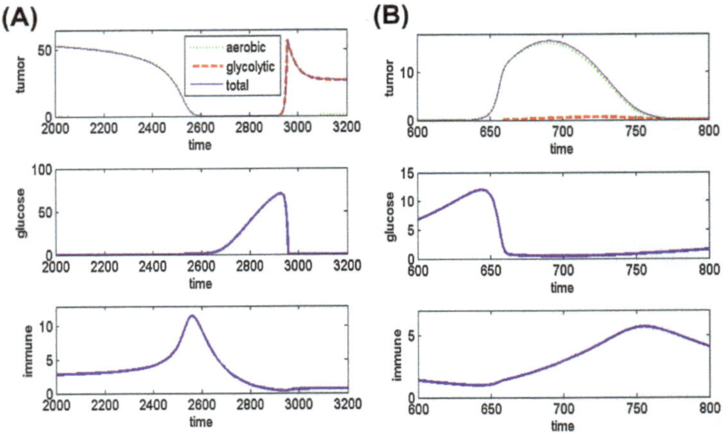

Figure 3.3 Details underlying the dynamics of (A) oscillations driven by glycolytic cancer cells ($b_T = 0.9$, $e_j = 0.1$, $r_g = 1.8r_a$, $i_0 = 2.1e - 4$) and (B) oscillations driven by aerobic cancer cells ($b_T = 0.2$, $e_j = 0.2$, $r_g = 2.8r_a$, $i_0 = 2.0e - 3$). The first case corresponds to competition-driven circumvention of immune surveillance; the second case corresponds to a variant of a classic predator–prey type system.

In such a scenario, the dynamics can be interpreted as *cooperation between aerobic and glycolytic tumor cells against cytotoxic lymphocytes*: glycolytic cells take up the resources, allowing the tumor to circumvent the immune system and for the more successful of the remaining tumor cells to grow (note that while in this case the immune system is circumvented by glycolytic cancer cells, the successful cancer cells could be either aerobic or glycolic).

WE WROTE AND ANALYZED THE MODEL. NOW WHAT?

Understanding the sequence of behaviors that are expected in such a system provides a framework for evaluating the effectiveness of the immune-stimulatory drugs. This understanding is important as the goal is to **take the dynamical behaviors out of the "unstable equilibrium" parameter range of pathological tumor growth to the "stable equilibrium" parameter range of tumor control and elimination**.

Inducing such a shift from one dynamical regime to the other can be done through application of various therapies, such as immunotherapy.

In recent years, significant advances have been made in the field of cancer immunotherapy. In particular, great strides have been made with immune checkpoint inhibitors, which target mechanisms that regulate immune cell activation and cytotoxic function against self-antigens. Alleviating some degree of checkpoint activity has been shown to significantly augment immune responses and improve outcomes in cancer patients, with particularly promising results seen in advanced melanoma patients [10]. However, excessive immune activation can lead to severe autoimmune responses, causing more harm than benefit. The risk of autoimmune effects can potentially be decreased with a more detailed understanding of both how a therapeutic agent acts once inside the body and how the body responds to it. These questions can be investigated using the tools of the well-developed field of pharmacokinetic and pharmacodynamic (PKPD) modeling. In what follows, we will provide a brief overview of how PKPD models are constructed, what information they can provide, and how classical academic mathematical modeling can supplement results of PKPD models.

PHARMACOKINETIC AND PHARMACODYNAMIC MODELING

Classic PKPD modeling involves (1) pharmacokinetic (PK) part of the process, which describes the process of drug administration, absorption, metabolism, and elimination and (2) pharmacodynamic (PD) part of the process, which describes the desired effect that the drug has on the body. Let us describe the two arms of PKPD in a little more detail, and see how our model can be used also suggest how academically developed mechanistic models (such as the one presented in this chapter) can be used to potentially elucidate the PD effects of the drug.

Drug Pharmacokinetics

PKPD modeling typically relies on the use of ordinary differential equations, which describe the rate of change of drug concentration as it travels through the body after administration. Depending on the mode of administration (oral, IV, subcutaneous, etc.), the drug may or may not go through an absorption phase. It then gets distributed into the tissue, where it binds to its target; the formation of the drug-target complex then hopefully leads to the desired therapeutic effect. At every stage, the body tries to eliminate the drug as a foreign agent, which is why it needs to be administered regularly to maintain the efficacious concentration of the agent in the body, and hence the desired effect.

The general form of PK equations that account for changes in concentrations of the drug in the periphery and at site of action can be described as follows:

$$\frac{dD_{\text{periphery}}}{dt} = \text{administration} - \text{absorption} - \text{elimination},$$

$$\frac{dD_{\text{site of action}}}{dt} = \text{absorption} - \text{binding to target} + \text{release from target} - \text{elimination}. \tag{3.2}$$

Because of all these forces acting on the therapeutic agent once it has been administered, it is important to evaluate the pharmacokinetic properties of the drug to estimate how much drug is available at the site of action at each moment. This is typically done through PK studies, where drug concentrations are measured at multiple time intervals after administration. The corresponding pharmacokinetic parameters (rates of absorption, elimination, volumes of distribution, etc.) are then estimated from the obtained data, often using very well-developed software, such as WinNonlin.

Detailed explanations of various aspects of PKPD modeling and the science behind it can be found in Ref. [11]. Numerous specific examples are available in the literature, see for instance [12,13].

Once the drug has reached the site of action and has been able to interact with the target, its therapeutic effects can be evaluated. In the field of immunooncology, models of tumor–immune interactions could provide very useful tools for evaluating the pharmacodynamic effects of immunotherapy from a mechanistic point of view.

Let us consider how such methods can be applied to the previously described model.

INCLUDING THE EFFECTS OF VARIOUS IMMUNOTHERAPEUTIC INTERVENTIONS

Many of the currently available immunotherapies act in one of the two main ways: through augmenting immune cell activation, such as using CTLA4 immune checkpoint inhibitors, or through decreasing immune cell mortality, such as through using PD-1 and PD-L1 immune checkpoint inhibitors; the mechanisms of action of these molecules are discussed in greater detail in the last chapter on therapeutic interventions.

Augmenting immune cell activation increases the value of parameter i_0, while decreasing immune cell mortality (potentially through delaying immune cell exhaustion) would decrease the value of the immune cell mortality parameter μ_I. Together, these two immunotherapies have the potential to augment immune cell responses to allow tumor elimination.

If drug concentration at the site of action is described by some function D, then its effects on augmenting immune cell stimulation can be accounted for in the following way in the equation for $I(t)$:

$$\frac{dI}{dt} = \underbrace{(i_0(T_a(t) + T_g(t)) * D}_{\text{increased immune stimulation due to drug activity}} - \underbrace{\mu_I I(t)}_{\text{natural decay}}) + \underbrace{r_I I(t)\left(\frac{G(t)}{\xi_I + b_I G(t)}\right)\left(\frac{I(t)(T_a(t) + T_g(t))}{I(t) + s(T_a(t) + T_g(t))}\right)}_{\text{cell expansion stimulated by debris from previously killed tumor cells and modulated by glucose}}$$

Increased Immune Stimulation Can Be Sufficient to Control Tumor Size but not to Eliminate It

Now, let us investigate the effects of further varying the value of parameter i_0 to simulate the effects of this type of immunotherapy. In our simulations in Fig. 3.4, we increased the value of i_0 by five orders of magnitude but found that it did not lead to tumor elimination. However, the increase did "destabilize" the tumor, yielding oscillatory regimes of a much smaller magnitude than in the previous set of simulations, thus moving it away from the stable equilibrium and into the "transitional regime" space, which is one step closer to tumor elimination.

One reason the immune cells fail to eliminate the tumor could be due to insufficient tumor recognition by the immune cells, which is accounted for by parameter $e_j, j = a, g$. We then hypothesized that even with strong supplementary immune stimulation, tumor elimination can be achieved only if the parameter of tumor elimination e_j is large enough. For this set of simulations, we fixed $i_0 = 2.1e - 4$ and then progressively increased parameter e_j (see Fig. 3.5).

These sets of simulations demonstrate that predictably, high enough rate of killing of tumor cells is necessary for tumor elimination. However, if parameter e_j cannot be manipulated or increased to a large enough value to completely eliminate the tumor, a higher degree of tumor control can still be achieved through sufficient immune stimulation. Mechanisms that augment immunogenicity are described in the last chapter of this book on therapeutic interventions. Hint: It is the therapeutic modality that is already in use, but administered at a different schedule and with different doses.

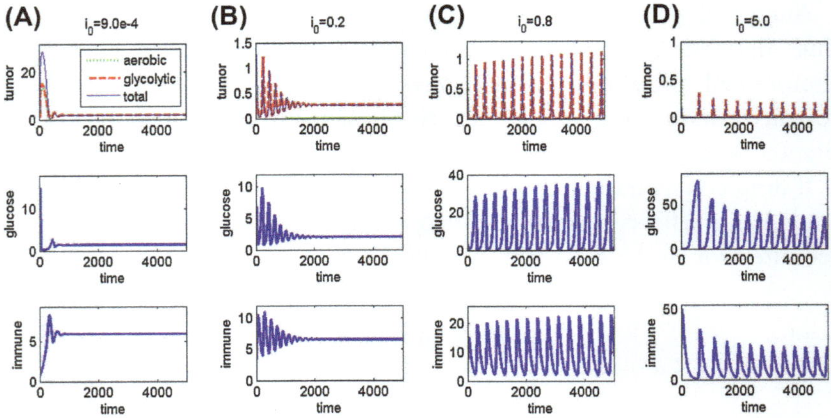

Figure 3.4 ***(A–D) Dynamical effects of increasing parameter*** $\mathbf{i_0}$**.** The simulations suggest that increased immune stimulation can decrease the amplitude of oscillations but cannot yield complete tumor elimination.

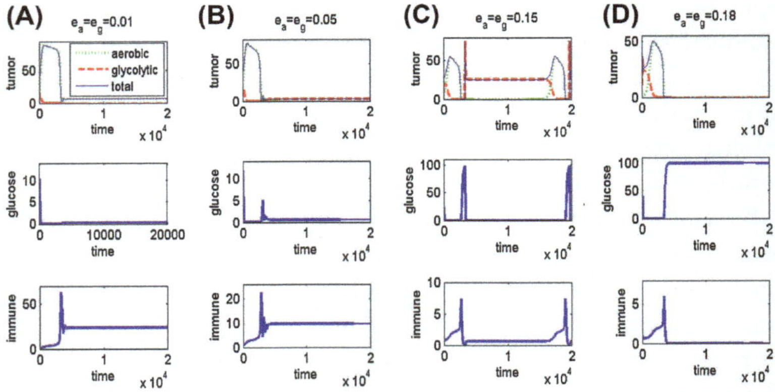

Figure 3.5 (A–D) Effects of increases of parameter of tumor elimination e_j, $j = a, g$ while keeping parameter $i_0 = 2.1e - 4$. Calculations suggest that sufficient increases in the value of e_j can indeed lead to full tumor elimination.

High Death Rates of Immune Cells Can Promote Tumor Growth

Another class of checkpoint inhibitors acts through decreasing immune cell mortality, such as blocking of the PD-1/PD-L1 connection. PD-1 is a receptor, whose expression increases on T cells when they reach a state of "exhaustion" [14]; at this point, "exhausted" T cells express PD-1, which can bind to the soluble ligand, such as PD-L1, which then signals termination of T cell life cycle. This mechanism exists to prevent the immune system from being active for too long and potentially harming the body.

Pharmacological blockade of either the PD-1 receptor or the corresponding PD-L1 ligand can help extend T cell life cycle, decreasing its mortality rate. Mathematically, this process can be captured by changing the value of the parameter representing natural immune cell mortality μ_I:

$$\frac{dI}{dt} = \underbrace{(i_0(T_a(t) + T_g(t))*D}_{\text{increased immune stimulation due to drug activity}} - \underbrace{(\mu_I - \mu_D * D)I(t)}_{\text{lower immune cell death due to immunotherapy}}) + \underbrace{r_I I(t)\left(\frac{G(t)}{\xi_I + b_I G(t)}\right)\left(\frac{I(t)(T_a(t) + T_g(t))}{I(t) + s(T_a(t) + T_g(t))}\right)}_{\text{cell expansion stimulated by debris from previously killed tumor cells and modulated by glucose}}$$

The decrease in immune cell mortality can be achieved via adding the term $\mu_D{}^*D^*I(t)$, provided that $\mu_I > \mu_D D$ (since we cannot have negative mortality).

To investigate the effects of variations in T cell mortality on the overall system dynamics, we conducted numerical experiments using parameter values that showed sustained tumor growth, and then observed how the effects of increases of parameter μ_I affect overall system dynamics. Parameters were taken to be $b_T = 0.9$, $i_0 = 0.0001$, $r_g = 2.8r_a$, $e_a = e_g = 0.1$; all other parameter values remained the same.

We observed that progressively decreasing the values of parameter μ_I (increasing immune cell mortality) leads to eventual tumor regression, preceded by progressively decreasing oscillatory regimes of small amplitude (see Fig. 3.6). These results suggest that, indeed, decreasing immune cell mortality can help achieve tumor control; however, one of course has to be careful about time and regulate these interventions to not cause severe immunotoxicities. Conversely, this also suggests that therapies that augment immune cell mortality (such as the standard protocol for chemotherapy administration) can undercut the positive effects of the therapy by catching immune cells in the cross fire. Therefore, in implementing these approaches, it is imperative to identify a healthy population range for immune cells $I(t)$ and maintain the schedule of drug administration D to keep immune cells in this range.

Hopefully, this example demonstrates how mathematical modeling can provide powerful tools that may help improve our understanding of the expected (and unexpected) effects of therapies. Combining PKPD modeling developed and employed in the industry with academically developed mechanistic modeling may help both augment our understanding of the biology and find more efficacious treatment approaches in a more fiscally responsible manner.

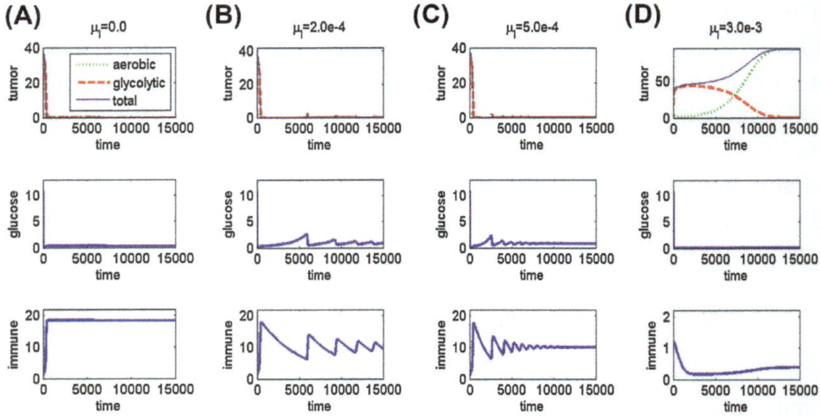

Figure 3.6 (A–D) This set of calculations illustrates the effect that increased mortality of immune cells can have on tumor dynamics, ranging from tumor escape (set a) to increased stabilizing oscillatory behavior to tumor elimination. Other parameters are held constant at $i_0 = 1.0e{-}4$, $b_T = 0.9$, $r_g = 2.8r_a$, $e_a = e_g = 0.1$.

GETTING BACK TO THE QUESTION

However, the question still remains: Can tumor-induced immune cell evasion explain why secondary tumors sometimes start growing after removal of primary tumors? In this chapter, we looked at tumor–immune interactions and how they can affect tumor growth. We showed that cancer and immune cells can compete for shared nutrients in the tumor microenvironment, and whoever gets to them first wins. (Notably, this idea of competition for shared nutrients between cancer and immune cells was recently experimentally validated by Chang et al. in [15], where the authors showed that using anti-PD-L1 checkpoint inhibitor interfered with anaerobic metabolism in tumors in mice, returning the competitive advantage to the immune cells and facilitating tumor regression).

Can this mechanism explain why metastases sometimes start growing faster after removal of primary tumors? It is possible that before surgery, some of the metastatic cancer cells were caught in cross-immunity, with the primary tumor stimulating systemic immune response that also kept metastatic microtumors at bay. Removing the primary tumor might relieve systemic inflammation, giving the secondary tumor an opportunity to build up a metabolically heterogeneous "shield" to initiate another competition-driven immunoediting cascade.

However, there do exist immune memory cells, whose half-lives are much longer than those of actively proliferating cytotoxic lymphocytes and who do not have such high nutrient demands. Furthermore, the ability of cancer cells to outcompete immune cells for nutrients is contingent (in this scenario) on the tumor's ability to grow large enough to form a glycolytic core, which would happen much further in the tumor progression timeline.

Therefore, while loss of cross-immunity is definitely a possibility, it is not a very likely one. It does however make us think of cancer as a disease that involves multiple systems of the body, each of which respond both to the disease and to each other.

In the next chapter, we will investigate another hypothesis, this time possibility having to do with blood vessel formation. I think we are getting closer to an answer.

REFERENCES

[1] Murphy K, Weaver C. Janeway's immunobiology. Garland Science; 2016.
[2] Gatenby RA, Gillies RJ. Why do cancers have high aerobic glycolysis? Nature Reviews Cancer. Nature Publishing Group 2004;4(11):891−9.
[3] Berg L, Tymoczko JM, Stryer JL. Biochemistry. 5th ed. New York: W.H. Freeman; 2002.
[4] Ganapathy V, Thangaraju M, Prasad PD. Nutrient transporters in cancer: relevance to Warburg hypothesis and beyond. Pharmacology & Therapeutics. Elsevier 2009;121(1):29−40.
[5] Fox CJ, Hammerman PS, Thompson CB. Fuel feeds function: energy metabolism and the T-cell response. Nature Reviews Immunology. Nature Publishing Group 2005;5(11):844−52.
[6] Kareva I, Berezovskaya F. Cancer immunoediting: a process driven by metabolic competition as a predator−prey−shared resource type model. Journal of Theoretical Biology. Elsevier 2015;380:463−72.
[7] Cantrell RS, Cosner C, Ruan S. Intraspecific interference and consumer-resource dynamics. Discrete and Continuous Dynamical Systems Series B. AIMS PRESS 2004;4:527−46.
[8] de Pillis LG, Radunskaya AE, Wiseman CL. A validated mathematical model of cell-mediated immune response to tumor growth. Cancer Research. AACR 2005;65(17):7950−8.
[9] de Pillis LG, Gu W, Radunskaya AE. Mixed immunotherapy and chemotherapy of tumors: modeling, applications and biological interpretations. Journal of Theoretical Biology. Elsevier 2006;238(4):841−62.
[10] Eggermont AM, Maio M, Robert C. Immune checkpoint inhibitors in melanoma provide the cornerstones for curative therapies. Seminars in Oncology. Elsevier 2015:429−35.
[11] Gabrielsson J, Weiner D. Pharmacokinetic and pharmacodynamic data analysis: concepts and applications. CRC Press; 2001.
[12] Chudasama VL, Zutshi A, Singh P, Abraham AK, Mager DE, Harrold JM. Simulations of site-specific target-mediated pharmacokinetic models for guiding the development of bispecific antibodies. Journal of Pharmacokinetics and Pharmacodynamics. Springer 2015;42(1):1−18.

[13] Singh AP, Krzyzanski W, Martin SW, Weber G, Betts A, Ahmad A, et al. Quantitative prediction of human pharmacokinetics for mAbs exhibiting target-mediated disposition. The AAPS Journal. Springer 2015;17(2):389–99.
[14] Wherry EJ. T cell exhaustion. Nature Immunology. Nature Publishing Group 2011;12(6):492–9.
[15] Chang C-H, Qiu J, O'Sullivan D, Buck MD, Noguchi T, Curtis JD, et al. Metabolic competition in the tumor microenvironment is a driver of cancer progression. Cell. Elsevier 2015;162(6):1229–41.

CHAPTER 4

Blood Vessel Formation and Pathological Angiogenesis as Mitigated by Competing Angiogenesis Regulators

Contents

INTRODUCTION

In 2006, several papers came out from Judah Folkman's lab, where the authors isolated several cancer cell lines that behaved in a very unusual fashion [1,2]. When these cells were grown in a Petri dish, they all grew at approximately the same rate. However, when these same cells were implanted in mice with compromised immune systems, some of them formed tumors within 30 days, some within 80, some after 120, and some never developed tumors at all. In each case, the tumors only grew once they were able to start forming their own blood vessels, a process that has become known as "angiogenic switch." The authors referred to the cell clones that underwent angiogenic switch and started growing relatively soon after implantation as "angiogenic clones," and the ones that grew much later, or not at all, as "nonangiogenic clones" (Fig. 4.1).

Understanding Cancer From a Systems Biology Point of View
ISBN 978-0-12-813673-7
http://dx.doi.org/10.1016/B978-0-12-813673-7.00004-5

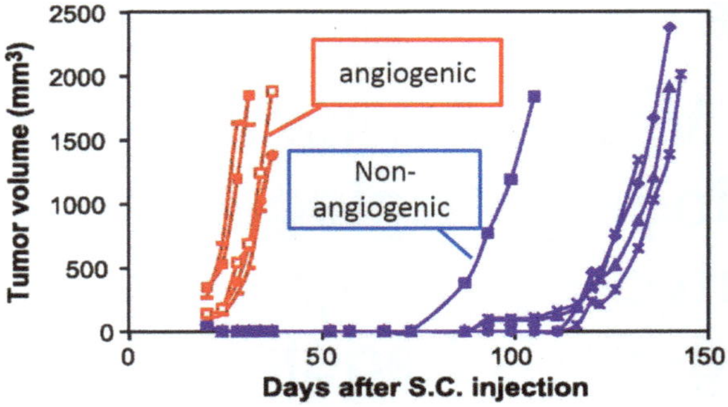

Figure 4.1 Angiogenic and nonangiogenic clones, grown in immunocompromised mice. *(The figure is adapted from Almog N, Henke V, Flores L, Hlatky L, Kung AL, Wright RD, et al. Prolonged dormancy of human liposarcoma is associated with impaired tumor angiogenesis. The FASEB Journal. FASEB 2006;20(7):947–49.)*

When they investigated these tumors closer, they observed that:

1. Both actively growing and temporarily dormant tumors had similarly shaped growth curves. In other words, once the tumors started growing, they behaved in the same way, with the only difference being the amount of time to angiogenic switch.
2. Prior to angiogenic switch, the only structures resembling vasculature in nonangiogenic clones were clusters of endothelial cells, with no lumens and no red blood cells. Therefore, these particular tumors were indeed incapable of growing until they could form their own blood vessels.
3. Balance between proliferation and apoptosis suggested that cells were not quiescent, and lack of growth was not a result of impairments in cell division or death rates.

These results were observed for various types of cancer, including liposarcoma, breast cancer, osteosarcoma, and glioblastoma. And of course, these observations lead to the question: What causes angiogenic switch? And why does it occur at different times for different tumors?

NORMAL BLOOD VESSEL FORMATION

The process of blood vessel formation is mediated by platelets, 2–5 μm subcellular structures formed by megakaryocytes in the bone marrow [3]. A typical adult with an average of 5 L of cardiac output would have 0.750–2.250 trillion circulating platelets at any given time. This fast turnover of large number of circulating platelets facilitates a very responsive system for regulating localized angiogenesis in an event of a wound.

As platelets circulate through the body, they pick up many angiogenesis-related proteins [4,5], which are stored in platelets in different compartments called alpha-granules [5]. In the event of tissue injury, platelets are recruited to the wound, where they become activated, undergoing morphological changes and forming extensions (lamellipodia and filipodia), which facilitate clotting. Platelet clots both physically block the injury and initiate formation of new blood vessels through sequential release of growth factors from alpha-granules.

Angiogenesis stimulators, such as vascular endothelial growth factor, or VEGF, induce formation of early vascular sprouts or tip cells. Sprout growth and lumenization (formation or stalk cells) are facilitated by factors, such as basic fibroblast growth factor, or bFGF, henceforth for brevity referred to as blood vessel stabilizers. Angiogenesis inhibitors, such as platelet factor 4 (PF-4), thrombospondin-1 (TSP-1), angiostatin or endostatin, facilitate termination of neovascularization by suppressing anti-apoptosis genes and inhibiting factors involved in cell cycle arrest, causing proliferating cells to become apoptotic. This causes vessel pruning and termination of the process of blood vessel formation [4,6–9].

After growth factors have been released from the platelet clot, they need to send out signals to initiate, promote, or terminate formation of new blood vessels, necessary for wound healing. However, to do so they need to first bind glycosaminoglycans (GAGs), such as heparan sulfate (HS), on the cells' surface [10,11]. The duration of signaling of each growth factor depends on its affinity GAGs, which in turn is determined by sensitivity to heparinase, an enzyme produced by fibroblasts in normal tissues and in the tumor microenvironment. Heparinase degrades GAGs, destroying the tissue anchor of the attached growth factor and causing it to detach from the binding site, effectively terminating its signaling. Affinity of each growth factor is determined by the amount of heparinase that is necessary to break the tissue anchor and cleave it.

Angiogenesis stimulators have the lowest affinity for heparinase, which means that it takes the lowest amount of heparinase to cleave the growth factor from the anchor.

Blood vessel stabilizers have higher comparative affinity for heparinase (medium affinity), which means that it requires higher amounts of heparinase to terminate the signaling.

Finally, angiogenesis inhibitors have the highest affinity for heparinase, indicating that it is very difficult to terminate inhibitive signaling once it has begun.

For as long as tissue injury exists, there is a constant upregulation of GAG formation. Once some healing occurs, HS is downregulated, scar tissue is deposited, and the system is turned off. Notably, the concentration of angiogenesis inhibitors in platelets has been reported to be orders of magnitude higher than that of stimulators [12], suggesting that the physiological state of angiogenesis in a healthy individual is "off."

The sequence of events underlying release of growth factors and subsequent effects on blood vessel formation are shown in Fig. 4.2.

WHAT HAPPENS IN TUMORS?

Many tumors are characterized by increased production of angiogenesis stimulators. Therefore, if there is a constant tumor-induced inflow of stimulators and stabilizers, they may cause angiogenesis inhibitors to become outcompeted for GAGs, effectively preventing termination of blood vessel formation.

One can think of angiogenesis regulators competing for the opportunity to attach to a GAG as **playing a game of musical chairs**. When inhibitors become outcompeted for their binding site "chairs," blood vessel formation does not stop, leading to continuous formation of leaky overgrowing vessels. Increased vessel growth provides tumor cells with more nutrients, which in turn allows further growth of cancer cells, creating a positive feedback loop.

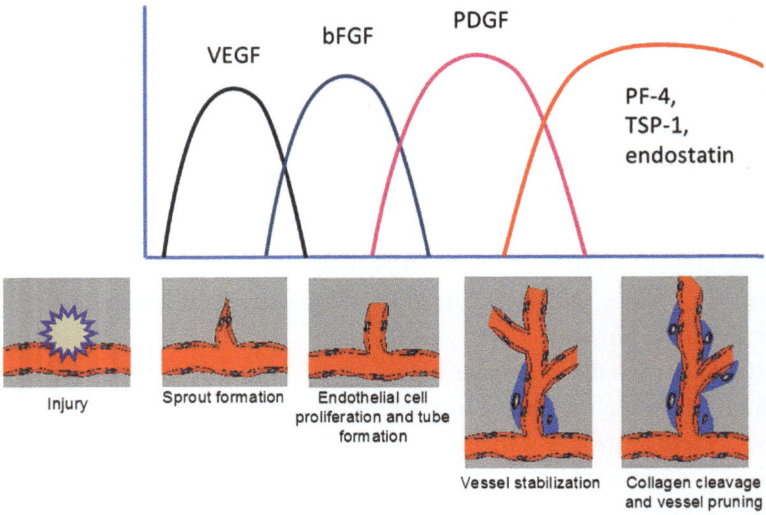

Figure 4.2 Sequential release of growth factors and corresponding steps in blood vessel formation. *(Figure adapted with permission from a presentation by Giannoula Klement.)*

USING MATHEMATICAL MODELING FOR HYPOTHESIS TESTING

Our first task in using mathematical modeling for hypothesis testing is to write a model that can quantitatively reproduce normal wound healing, showing vessel growth, stabilization, and pruning. Next, after we have thus validated our model, we can use it to test our hypothesis and to show that:

1. Increased inflow of angiogenesis regulators can indeed cause vasculature buildup.
2. This buildup can lead to tumor escape.
3. Variations in time to escape can be reproduced with varying just the parameters that determine the degree to which the tumor stimulates its environment.

And if we can achieve these goals, we can get back to the main question:

4. Could the tumor–stroma interactions be the reason why secondary tumors show accelerated growth after the primary tumor is removed?

Let us tackle these one by one.

MODEL DESCRIPTION

To show that increased inflow of regulators can cause buildup of vasculature, we created the following model [13], which takes into account the dynamical activity over time of the following variables: tumor $T(t)$; vasculature $V(t)$; unbound low- (e.g., VEGF), medium- (e.g., bFGF), and high-affinity (e.g., PF-4) growth factors $L_u(t)$, $M_u(t)$, and $H_u(t)$; bound low-, medium-, and high-affinity growth factors $L_b(t)$, $M_b(t)$, and $H_b(t)$; and free binding sites $B(t)$.

The change in volume of tumor $T(t)$ is described by a logistic growth function with a dynamic carrying capacity $\frac{dT}{dt} = \lambda T(t)\left(1 - \frac{T(t)}{k+\varepsilon V(t)}\right)$, where λ is the growth rate of the tumor cells, k is tumor carrying capacity in the absence of neovascularization (the maximum size that a dormant nonvascularized tumor can reach), ε is a scaling constant and $V(t)$ is the volume of new vasculature, formed as a result of the activity of angiogenesis stimulators and inhibitors.

The simulation begins with the release of unbound angiogenesis regulators $L_u(t)$, $M_u(t)$, and $H_u(t)$ from the platelet clot at the site of the wound or the tumor. Unbound regulators attach to the free binding sites $B(t)$ at different rates. Bound growth factors eventually get discarded from the system at different rates: growth factors with lower affinity detach at a higher rate, while growth factors with higher affinity detach at lower rates. Once growth factors are detached, the binding sites are released and become available for other growth factors that might be competing for them. In the presence of a tumor $T(t)$, there might be additional inflow of unbound angiogenesis regulators. This is accounted for by the terms $\alpha T(t)$, $\beta T(t)$, and $\gamma T(t)$ for the three types of growth factors, respectively.

The resulting equations describing the dynamics of the activity of unbound and bound growth factors are thus as follows:

$$\underbrace{\frac{dL_u(t)}{dt}}_{\substack{\text{unbound low}\\ \text{affinity cytokines}\\ \text{(e.g., VEGF)}}} = \underbrace{-aL_u(t)B(t)}_{\substack{\text{rate of attachment}\\ \text{of } L_u \text{ to binding cites}}} + \underbrace{\alpha T(t)}_{\substack{\text{tumor-induced}\\ L_u \text{ inflow}}}, \quad \underbrace{\frac{dL_b(t)}{dt}}_{\substack{\text{bound low}\\ \text{affinity cytokines}\\ \text{(e.g., VEGF)}}} = \underbrace{aL_u(t)B(t)}_{\substack{\text{rate of inflow due}\\ \text{to } L_u \text{ attachment}\\ \text{to binding cites}}} - \underbrace{bL_b(t)}_{\text{discarded } L_b}$$

$$\frac{dM_u(t)}{dt} = -cM_u(t)B(t) + \beta T(t), \qquad \frac{dM_b(t)}{dt} = cM_u(t)B(t) - dM_b(t)$$

$$\frac{dH_u(t)}{dt} = -eH_u(t)B(t) + \gamma T(t), \qquad \frac{dH_b(t)}{dt} = eH_u(t)B(t) - fH_b$$

The number of the available binding sites $B(t)$ decreases when they become occupied by unbound angiogenesis regulators $L_u(t)$, $M_u(t)$, and $H_u(t)$ and increases when bound angiogenesis regulators $L_b(t)$, $M_b(t)$, and $H_b(t)$ detach and are discarded. We also introduce a logistic saturation term $\left(1 - \frac{B(t)}{K+\xi T(t)}\right)$ to incorporate carrying capacity K for the number of binding sites that can become available during normal blood vessel formation. In the presence of the tumor, the advancing tumor edge will increase this carrying capacity by introducing more cells with more binding sites on them. This effect is accounted for in the model by the denominator $K + \xi T(t)$. The dynamic carrying capacity does not limit the process of attachment once the binding sites are available, but it does control, whether the released binding sites will become available for another use. These considerations are incorporated in the following equation:

$$\underbrace{\frac{dB(t)}{dt}}_{\substack{\text{available}\\ \text{binding cites}}} = \underbrace{(bL_b(t) + dM_b(t) + fH_b(t))}_{\substack{\text{replenishment of number of B(t) after they}\\ \text{are released from being occupied by } L_b, M_b, H_b}} \underbrace{\left(1 - \frac{B(t)}{K + \xi T(t)}\right)}_{\text{saturation term}} - \underbrace{B(t)(aL_u(t) + cM_u(t) + eH_u(t))}_{\substack{\text{reduction in number of free binding cites}\\ \text{due to occupation by } L_u, M_u \text{ and } H_u}}.$$

The increase in volume of new vasculature $V(t)$ occurs due to the signaling from bound angiogenesis stimulators $L_b(t)$ and blood vessel stabilizers $M_b(t)$ and is terminated by the bound angiogenesis inhibitors $H_b(t)$. We describe increase in neovascularization by the product of $L_b(t)$ and $M_b(t)$ rather than by their sum because functional blood vessels require synergistic signaling of both angiogenesis stimulators and blood vessel stabilizers. The spatial limitations for neovasculature volume are accounted for by the carrying capacity V_{max}. The dynamics of $V(t)$ is therefore described by the following equation:

$$\underbrace{\frac{dV}{dt}}_{\text{new blood vessels}} = \underbrace{\delta L_b(t) M_b(t)}_{\substack{\text{increase due to activity of}\\ \text{low and medium affinity}\\ \text{grwoth factors}\\ \text{(e.g., VEGF and bFGF)}}} \underbrace{\left(\eta - V(t)\frac{H_b(t)}{V_{max}}\right)}_{\substack{\text{decrease due to activity}\\ \text{of high affinity growth}\\ \text{factors (e.g., PF}-4\text{),}\\ \text{with total volume capacity } V_{max}}}.$$

The final system of equations is thus as follows:

$$\underbrace{\frac{dL_u(t)}{dt}}_{\substack{\text{unbound low}\\ \text{affinity cytokines}\\ \text{(e.g., VEGF)}}} = \underbrace{-aL_u(t)B(t)}_{\substack{\text{rate of attachment}\\ \text{to binding cites}}} + \underbrace{\alpha T(t)}_{\substack{\text{tumor}-\text{induced}\\ \text{VEGF inflow}}} \qquad \underbrace{\frac{dL_b(t)}{dt}}_{\substack{\text{bound low}\\ \text{affinity cytokines}\\ \text{(e.g., VEGF)}}} = \underbrace{aL_u(t)B(t)}_{\substack{\text{rate of inflow due}\\ \text{to } L_u \text{ attachment}\\ \text{to binding cites}}} - \underbrace{bL_b(t)}_{\text{discarded } L_b}$$

$$\frac{dM_u(t)}{dt} = -cM_u(t)B(t) + \beta T(t) \qquad \frac{dM_b(t)}{dt} = cM_u(t)B(t) - dM_b(t)$$

$$\frac{dH_u(t)}{dt} = -eH_u(t)B(t) + \gamma T(t) \qquad \frac{dH_b}{dt} = eH_u(t)B(t) - fH_b$$

$$\underbrace{\frac{dB(t)}{dt}}_{\substack{\text{available}\\ \text{binding cites}}} = \underbrace{(bL_b(t) + dM_b(t) + fH_b(t))}_{\substack{\text{replenishment of number of } B(t)\\ \text{after they are released}}} \underbrace{\left(1 - \frac{B(t)}{K + \xi T(t)}\right)}_{\text{saturation term}} \underbrace{- B(t)(aL_u(t) + cM_u(t) + eH_u(t))}_{\substack{\text{reduction in number of free binding cites}\\ \text{due to occupation by } L_u, M_u \text{ and } H_u}}$$

$$\underbrace{\frac{dV}{dt}}_{\text{new blood vessels}} = \underbrace{\delta L_b(t) M_b(t)}_{\substack{\text{increase due to activity}\\ \text{of low}-\text{ and medium}-\\ \text{affinity factors}\\ \text{(e.g., VEGF and PDGF)}}} \underbrace{\left(\eta - V(t)\frac{H_b(t)}{V_{\max}}\right)}_{\substack{\text{decrease due to activity of}\\ \text{high}-\text{affinity growth}\\ \text{factors (e.g., PF}-4\text{), with}\\ \text{total volume capacity } V_{\max}}}$$

$$\underbrace{\frac{dT}{dt}}_{\text{tumor growth rate}} = \underbrace{\lambda T(t)\left(1 - \frac{T(t)}{k + \varepsilon V(t)}\right)}_{\substack{\text{logistic growth rate, with carrying capacity}\\ \text{determined by new vasculature}}} \tag{4.1}$$

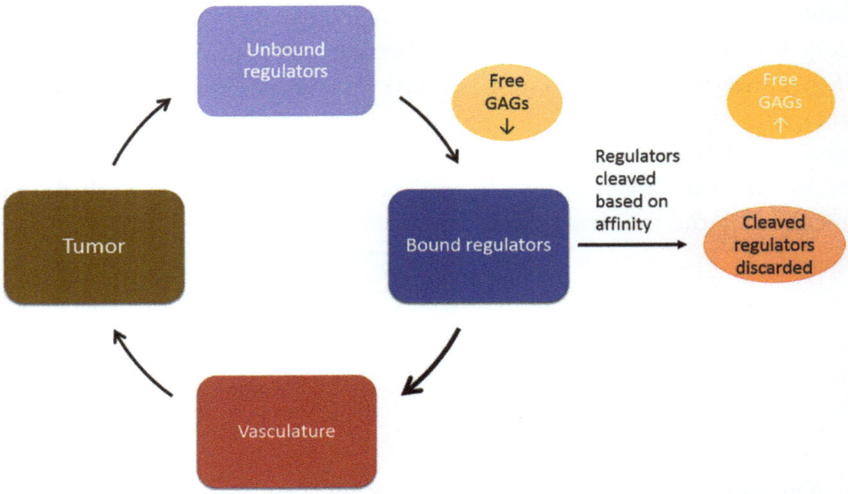

Figure 4.3 Schematic representation of the mechanism described in System (4.1).

The schematic mechanism of activity of angiogenesis regulators as they attach and detach from binding sites is shown in Fig. 4.3.

PARAMETER ESTIMATION

Since the proposed model is conceptual in nature, no specific parameter set exists, from which values can be comprehensively derived. However, a number of parameter values can be estimated based on previously published models and on theoretical considerations. In a series of in vitro experiments, Kareva et al. [4] demonstrated that growth factors are released from platelet clots in a sequence that depends on each growth factor's affinity for GAGs, and particularly heparan sulfate. Measuring exact rates of attachment and detachment of angiogenesis regulators is very challenging due to the transient nature of these events. However, it is possible to estimate relative affinities of these parameters, which allows us to say that $a > c > e$ and $b > d > f$. Parameter for tumor growth rate λ was chosen based on values reported in Ref. [14]. Parameter k represents the maximum size that a dormant nonvascularized tumor can attain, which is estimated to be ≈ 1 mm in diameter. The value of parameter K, which represents carrying capacity for the number of available binding sites, is expected to vary from tissue to tissue and was chosen arbitrarily, as was parameter ξ. Parameters α,

β, and γ, which represent the degree of tumor-induced stromal stimulation, will be the focus of the current investigation and are therefore varied, while other parameters remain fixed unless otherwise indicated. The summary and sample values of the parameters used are given in Table 4.1. Relative importance of these parameters will be investigated in the next section using the Fourier Amplitude Sensitivity Test (FAST) method.

NORMAL BLOOD VESSEL FORMATION

First, we will use the model to reproduce a pattern of behavior consistent with normal blood vessel formation. In this case, there exists no additional inflow of angiogenesis regulators beyond those that were initially released from platelets, so parameters α, β, and γ are taken to be zero. As can be seen in Fig. 4.4A, attachment of growth factors to the binding sites occurs in a sequential manner, with angiogenesis stimulators with lower affinity for HS binding first, followed by stabilizers, and finally by inhibitors. The corresponding effect on vascular formation can be seen in Fig. 4.4B, where volume of new blood vessels first increases, and then decreases due to inhibitor-induced vasculature pruning. These dynamics are consistent with the expected pattern of normal neovascularization.

The Effects of Tumor-Induced Inflow of Angiogenesis Regulators

Next, we change values of parameters α, β, and γ to simulate the effects of tumor-induced inflow of angiogenesis regulators on vasculature formation and corresponding tumor growth.

First, we investigated a scenario, where a simulated microtumor induces production of just angiogenesis stimulators. We do so by setting $\alpha = 3$, $\beta = \gamma = 0$, $T(0) = 0.1$, and the rest of parameters were taken as in Table 4.1. As can be seen from Fig. 4.5A–C, while the predicted cytokine profile of bound angiogenesis regulators differs from normal (Fig. 4.4A), additional inflow of just stimulators is insufficient to cause escape from a steady state of tumor dormancy. This is physiologically plausible, since as was mentioned above, in nonangiogenic dormant tumors, the only structures resembling vasculature that have been experimentally observed are clusters of endothelial cells, with no lumens and no red blood cells [1,15]. To secure formation of lumens, signaling from medium-affinity growth factors would be required.

Table 4.1 Brief description, estimated values, and units of variables and parameters introduced in System (4.1)

	Description	Estimated value	Units
$L_u(t)$ and $L_b(t)$	Unbound and bound low-affinity angiogenesis regulators, (e.g., VEGF, a stimulator of angiogenesis)	$L_u(0) > 0 \;\; L_b(0) = 0$	vol
$M_u(t)$ and $M_b(t)$	Unbound and bound medium-affinity angiogenesis regulators (e.g., bFGF, a vessel stabilizing factor)	$M_u(0) > 0 \;\; M_b(0) = 0$	vol
$H_u(t)$ and $H_b(t)$	Unbound and bound high-affinity angiogenesis regulators (e.g., PF-4, TSP-1, and endostatin, angiogenesis inhibitors)	$H_u(0) > 0 \;\; H_b(0) = 0$	vol
$B(t)$	Available binding sites for angiogenesis regulators	$B(t) > 0$	vol
$V(t)$	New vasculature	$V(t) \geq 0$	vol
$T(t)$	Tumor volume	$T(t) \geq 0$	vol (mm^3)
a	Rate at which unbound $L_u(t)$ become bound	$a > c > e > 0 \;\; a = 0.29$	$\frac{1}{\text{time vol}}$
b	Rate of detachment of $L_b(t)$ from binding sites $B(t)$	$b > d > f > 0 \;\; b = 0.5$	$\frac{1}{\text{time}}$
c	Rate at which unbound $M_u(t)$ become bound	$a > c > e > 0 \;\; c = 0.1$	$\frac{1}{\text{time vol}}$
d	Rate of detachment of $M_b(t)$ from binding sites $B(t)$	$b > d > f > 0 \;\; d = 0.1$	$\frac{1}{time}$
e	Rate at which unbound $H_u(t)$ become bound	$a > c > e > 0 \;\; e = 0.03$	$\frac{1}{\text{time vol}}$
f	Rate of detachment of $H_b(t)$ from binding sites $B(t)$	$b > d > f > 0 \;\; f = 0.01$	$\frac{1}{time}$
K	Carrying capacity for the number of possible available binding sites	$K = 10$	vol
ξ	Steepness coefficient for tumor-induced expansion of the carrying capacity for the number of available binding sites	$\xi = 5 \times 10^2$	Unitless

δ	Rate of expansion of neovasculature due to signaling of $L_b(t)$ and $M_b(t)$	$\delta = 0.05$	$\frac{1}{\text{time vol}^2}$
η	Scaling parameter for increase in neovascularization due to activity of $L_b(t)$ and $M_b(t)$	$\eta = 1$	vol
V_{max}	Carrying capacity for neovasculature	$V_{max} > 0$ $V_{max} = 10$	vol
λ	Tumor growth rate	$\lambda = 0.04$	$\frac{1}{\text{time}}$
k	Carrying capacity of a nonvascularized dormant tumor	$k = 0.1$	vol (mm^3)
ε	Scaling parameter for increase of neovasculature as tumor's dynamic carrying capacity	$\varepsilon = 0.65$	Unitless
α	Rate of tumor-induced inflow of angiogenesis stimulators $L_u(t)$	$\alpha \geq 0$	$\frac{1}{\text{time}}$
β	Rate of tumor-induced inflow of blood vessel stabilizers $M_u(t)$	$\beta \geq 0$	$\frac{1}{\text{time}}$
γ	Rate of tumor-induced inflow of angiogenesis inhibitors $H_u(t)$	$\gamma \geq 0$	$\frac{1}{\text{time}}$

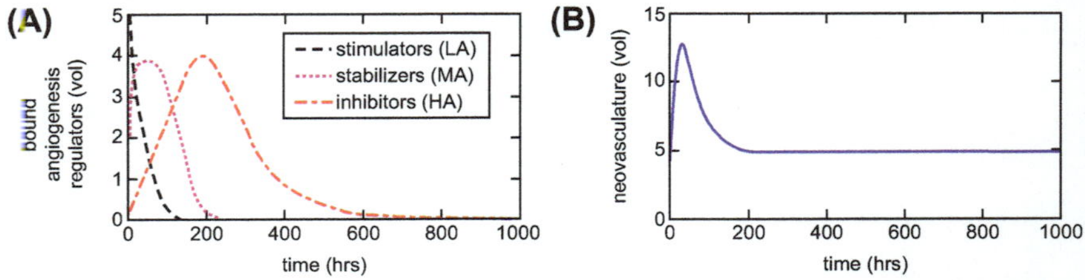

Figure 4.4 *Neovasculature formation in response to signaling of angiogenesis regulators.* (A) The pattern of activity of angiogenesis regulators after they have been released from the platelet clot and have bound to heparan sulfate (HS). First we observe attachment and release of angiogenesis stimulators (e.g., VEGF) with lowest affinity for HS, followed by blood vessel stabilizers (e.g., bFGF) with medium affinity for HS, followed by angiogenesis inhibitors (e.g., PF-4) with highest affinity for HS. (B) The pattern of neovasculature formation that occurs in response to the activity of angiogenesis regulators. During the spike of activity of stimulators LA and stabilizers MA, we observe an increase in neovasculature volume, followed by pruning in response to the activity of inhibitors HA. This behavior is consistent with normal pattern of neovascularization. *(This figure is used with permission from Kareva I. Escape from tumor dormancy and time to angiogenic switch as mitigated by tumor-induced stimulation of stroma. Journal of Theoretical Biology Elsevier 2016;395:11–22.)*

Next, we introduce tumor-induced inflow of both angiogenesis stimulators and blood vessel stabilizers by setting $\alpha = 0.2$, $\beta = 0.15$, and keeping the rest of the parameters constant. In this case, after a delay, we could observe escape from dormancy (Fig. 4.5D–F). According to the results of these simulations, in order for sufficient vasculature to develop so that the tumor can escape dormancy, it has to induce production of both angiogenesis stimulators and blood vessel stabilizing factors.

Next, we increase the value of parameters $\alpha = 0.35$, $\beta = 0.55$ to investigate, whether the time to escape can be accelerated or delayed through modification of parameters of the degree of stromal stimulation. Indeed, while in the previous case, the escape occurred at $t \approx 1500$ h, which is equivalent to 62.5 days (Fig. 4.5F), now escape occurs at $t \approx 600$ h, which is equivalent to 25 days.

Finally, we investigate a scenario, where $\alpha = 0.35$, $\beta = 0.55$, and $\gamma = 0.1$, that is, a case when there exists a tumor-induced production of all types of angiogenesis regulators, including stimulators, stabilizers, and inhibitors. The model predicts that if there exists a sufficient inflow of angiogenesis inhibitors, it can lead to vascular normalization and prevent escape from dormancy.

Therefore, the larger the values of α and β, the faster the process of neovascularization can occur.

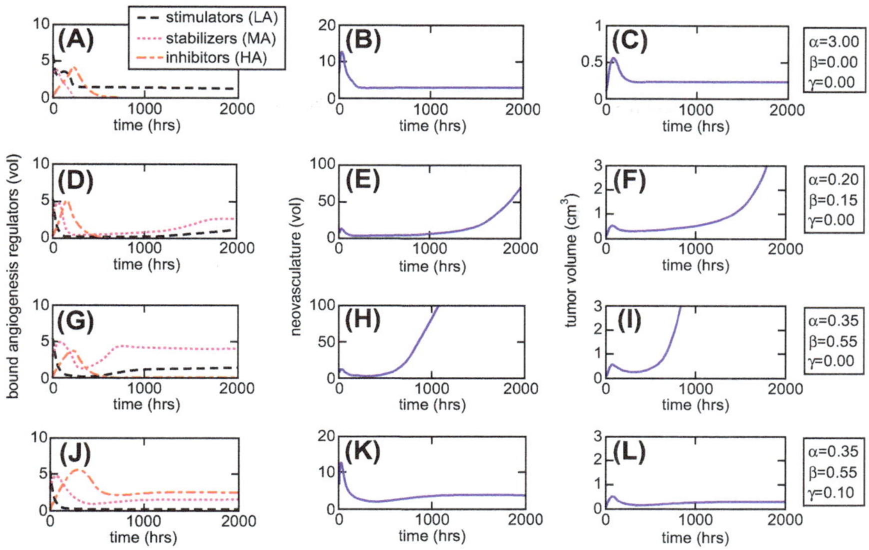

Figure 4.5 ***Effects of tumor-induced inflow of angiogenesis regulators on predicted cytokine profile of bound angiogenesis regulators, on neovasculature, and on tumor size.*** Varied parameters that are α, which represents tumor-induced stimulation of production of angiogenesis stimulators (e.g., VEGF); β, which represents tumor-induced stimulation of production of blood vessel stabilizers (e.g., bFGF); γ, which represents tumor-induced stimulation of production of angiogenesis inhibitors (e.g., PF-4). All the other parameter values are taken from Table 4.1. For mild tumor-induced stromal stimulation, given by $\alpha = 3$ (tumor-induced stimulation of production of just VEGF), we observe a close to normal pattern of activity of bound angiogenesis regulators (A), normal vasculature formation (B), and no escape from tumor dormancy (C). Tumor-induced stimulation of production of both angiogenesis stimulators and stabilizers, given by $\alpha = 0.2$, $\beta = 0.15$ resulted in changes in the activity of angiogenesis regulators (D), increase in neovasculature after $t = 1500$ h (62.5 days), which can be interpreted as angiogenic switch (E, note y-axis values), and tumor escape (F). Higher degree of tumor-induced stromal stimulation given by $\alpha = 0.35$, $\beta = 0.55$ resulted in even more abnormal pattern of activity of angiogenesis regulators (G), faster time to angiogenic switch at $t = 600$ h, or 25 days (H, note y-axis values) and faster time to escape from dormancy (I). However, additional inflow of angiogenesis inhibitors $\alpha = 0.35, \beta = 0.55, \gamma = 0.1$ is predicted to stabilize pattern of activity of angiogenesis regulators (J), stabilizing neovasculature (K), and preventing tumor escape (L). *(This figure is used with permission from Kareva I. Escape from tumor dormancy and time to angiogenic switch as mitigated by tumor-induced stimulation of stroma. Journal of Theoretical Biology. Elsevier 2016;395:11–22.)*

Comparison With Previously Reported Experimental Data

Next, we want to compare predictions of this model with experimentally obtained growth curves of both nonangiogenic dormant and angiogenic actively growing tumors.

In previous sections, we hypothesized and then showed that variations in values of parameters α, β, and γ can in fact lead to variations in time to angiogenic switch. Indeed, as one can see in Fig. 4.6, we can qualitatively reproduce experimentally observed growth curves with the proposed mathematical model solely through variations of these parameters.

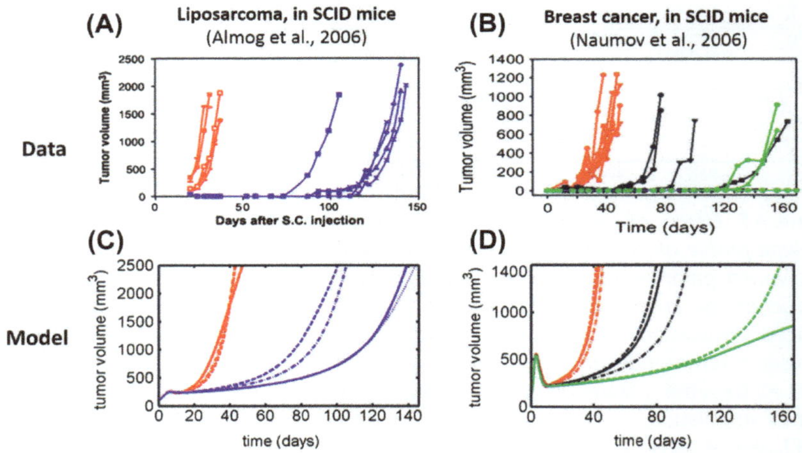

Figure 4.6 ***Comparison of model predictions with previously reported experimental data.*** The authors isolated and reported growth curves for nonangiogenic dormant and angiogenic actively growing tumor clones that were implanted in SCID mice. Qualitatively similar growth curves we reproduced with mathematical model described in System (4.1) using variations of parameters α, β, and γ, which represent tumor-induced stimulation of stroma to produce angiogenesis stimulators, blood vessel stabilizers, and angiogenesis inhibitors (B). Qualitatively similar growth curves were once again reproduced with the proposed mathematical model using just the variations in parameters α, β, and γ. Notably, different combinations of these parameters can reproduce qualitatively similar growth curves, suggesting that tumors capable of stromal stimulation of different growth factors can produce qualitatively similar growth curves with similar time to angiogenic switch. *(The data in (A) were reported in Almog N, et al. Prolonged dormancy of human liposarcoma is associated with impaired tumor angiogenesis. FASEB Journal 2006;20:947–49. (B) Naumov GN, et al. A model of human tumor dormancy: an angiogenic switch from the nonangiogenic phenotype, Journal of the National Cancer Institute 2006;98:316–25. This figure is modified with permission from Kareva I. Escape from tumor dormancy and time to angiogenic switch as mitigated by tumor-induced stimulation of stroma. Journal of Theoretical Biology. Elsevier 2016;395:11–22.)*

Moreover, different growth curves that follow a similar trajectory can be produced with variations of combinations of values of α, β, and γ, which can be either tumor-specific or tissue-specific. That is, for instance, a curve that corresponds to escape of a breast cancer microtumor can be obtained by taking $\{\alpha = 0.3, \beta = 0.3\}$ or by taking $\{\alpha = 0.2, \beta = 0.4\}$, as can be seen in Fig. 4.6D.

These results suggest that the degree of stromal stimulation by the tumor can be a sufficient mechanism to explain the delay in tumor growth and the onset of the angiogenic switch. It is a very exciting result, because it highlights how a very complicated mechanism of blood vessel formation can be broken down into mechanisms that can be described by just a few equations. And even more importantly, this model provides a mechanistic explanation for a phenomenon of "angiogenic switch" without requiring additional assumptions about specific mutations in cancer cells other than those that pertain to tumor-induced production of angiogenesis regulators.

SUMMARY

In this chapter we used a mathematical model that allows us to show that "in a world, where our assumptions are true," (1) increased inflow of angiogenesis regulators from the tumor can indeed cause a "buildup" of vasculature, (2) this buildup can lead to tumor escape, and (3) experimentally observed variations in time to escape can be reproduced with varying just the parameters that determine the degree to which the tumor stimulates the stroma.

Now, could this mechanism underlie the answer to our main question?

REFERENCES

[1] Naumov GN, Bender E, Zurakowski D, Kang S-Y, Sampson D, Flynn E, et al. A model of human tumor dormancy: an angiogenic switch from the nonangiogenic phenotype. Journal of the National Cancer Institute. Oxford University Press 2006;98(5):316–25.

[2] Almog N, Henke V, Flores L, Hlatky L, Kung AL, Wright RD, et al. Prolonged dormancy of human liposarcoma is associated with impaired tumor angiogenesis. The FASEB Journal. FASEB 2006;20(7):947–9.

[3] Italiano J, Shivdasani R. Megakaryocytes and beyond: the birth of platelets. Journal of Thrombosis and Haemostasis. Wiley Online Library 2003;1(6):1174–82.

[4] Kareva I, Abou-Slaybi A, Dodd O, Dashevsky O, Klement GL. Normal wound healing and tumor angiogenesis as a game of competitive inhibition. PLoS One 2016;11(12):e0166655. http://dx.doi.org/10.1371/journal.pone.0166655.

[5] Italiano JE, Richardson JL, Patel-Hett S, Battinelli E, Zaslavsky A, Short S, et al. Angiogenesis is regulated by a novel mechanism: pro-and antiangiogenic proteins are

organized into separate platelet α granules and differentially released. Blood. American Society of Hematology 2008;111(3):1227–33.
[6] Hellström M, Phng L-K, Hofmann JJ, Wallgard E, Coultas L, Lindblom P, et al. Dll4 signalling through Notch1 regulates formation of tip cells during angiogenesis. Nature. Nature Publishing Group 2007;445(7129):776–80.
[7] Gerhardt H. VEGF and endothelial guidance in angiogenic sprouting. VEGF in Development. Springer; 2008. p. 68–78.
[8] Blanco R, Gerhardt H. VEGF and Notch in tip and stalk cell selection. Cold Spring Harbor Perspectives in Medicine. Cold Spring Harbor Laboratory Press 2013;3(1):a006569.
[9] Xie L, Duncan MB, Pahler J, Sugimoto H, Martino M, Lively J, et al. Counter-balancing angiogenic regulatory factors control the rate of cancer progression and survival in a stage-specific manner. Proceedings of the National Academy of Sciences. National Academy Sciences 2011;108(24):9939–44.
[10] Vlodavsky I, Eldor A, Bar-Ner M, Fridman R, Cohen IR, Klagsbrun M. Heparan sulfate degradation in tumor cell invasion and angiogenesis. Advances in Experimental Medicine and Biology 1988;233:201.
[11] Vlodavsky I, Miao H, Atzmon R, Levi E, Zimmermann J, Bar-Shavit R, et al. Control of cell proliferation by heparan sulfate and heparin-binding growth factors. Thrombosis and Haemostasis 1995;74(1):534.
[12] Peterson JE, Zurakowski D, Italiano JE, Michel LV, Fox L, Klement GL, et al. Normal ranges of angiogenesis regulatory proteins in human platelets. American Journal of Hematology. Wiley Online Library 2010;85(7):487–93.
[13] Kareva I. Escape from tumor dormancy and time to angiogenic switch as mitigated by tumor-induced stimulation of stroma. Journal of Theoretical Biology. Elsevier 2016;395:11–22.
[14] de Pillis LG, Radunskaya AE, Wiseman CL. A validated mathematical model of cell-mediated immune response to tumor growth. Cancer Research. AACR 2005;65(17):7950–8.
[15] Naumov GN, Akslen LA, Folkman J. Role of angiogenesis in human tumor dormancy: animal models of the angiogenic switch. Cell Cycle. Taylor & Francis 2006;5(16):1779–87.

CHAPTER 5

Angiogenesis Regulators as a Possible Key to Accelerated Growth of Secondary Tumors Following Primary Tumor Resection

Contents

INTRODUCTION

Metastases are a primary cause of death from cancer, and as was discussed throughout this book, numerous observations suggest that surgical removal of primary tumors can sometimes lead to accellerated growth of secondary tumors [1–5]. These observations suggest that primary tumors may have an inhibitory effect on metastases. This phenomenon has been observed in colorectal cancer [4] and non–small cell lung cancer [6], among other cancer types.

A number of hypotheses have been proposed to explain this observation. Some of these hypotheses include (1) monopolization of resources by the primary tumor, which deprives secondary tumors of necessary nutrients; (2) immune activation induced by the primary tumor, which causes cross-immunity and thus suppression of secondary tumors [7–9]; (3) some form of intrinsic population heterogeneity, which can dictate time to escape from dormancy regardless of extrinsic events [10], among others.

Understanding Cancer From a Systems Biology Point of View
ISBN 978-0-12-813673-7
http://dx.doi.org/10.1016/B978-0-12-813673-7.00005-7

In the previous chapters, we explored the possibility that growth of dormant tumors, whether primary or secondary, is governed by intrinsic growth laws. We concluded that this mechanism is not very likely, since removal of primary tumors would have no real effect on independently growing secondary tumors.

We also investigated the possibility that cancer cells outcompete immune cells for nutrients, thereby creating an immunosuppressive environment. However, this mechanism can come into effect only once the tumor has had an opportunity to grow and create hypoxic regions to truly disarm the immune system. Therefore, this mechanism is also not a likely candidate.

Finally, we looked at the process of neovascularization, specifically, how growth factors released from platelets compete for binding sites during both normal and pathological angiogenesis. We showed that tumor-induced production of angiogenesis stimulators and blood vessel stabilizers outcompete angiogenic inhibitors in the tumor microenvironment, creating a positive feedback loop that permits self-supporting tumor growth. Through varying parameters governing the degree of tumor-induced production of angiogenesis regulators, we were able to qualitatively replicate experimentally observed growth curves for both dormant and actively growing liposarcoma and breast cancer. In fact, varying only two parameters was sufficient to replicate any experimentally observed time to angiogenic switch in the available data. The model thus provided a possible mechanistic explanation to escape from dormancy of primary tumors.

Here, we take the discussion of this mechanism a step further and focus on the systemic response of the host to tumor-induced production of angiogenesis stimulators, and on the long-term effects of the altered balance between stimulators and inhibitors in the presence of more than one tumor in the host's body.

As was mentioned before, many tumors are characterized by increased production of angiogenesis stimulators, such as vascular endothelial growth factor (VEGF) [11–13]. Furthermore, it has been observed that the level of inhibitors in the tumor host is also elevated, production of which has been attributed to the tumors themselves, suggesting a degree of self-inhibition [14–16]. This hypothesis was mathematically investigated by Benzekry et al. [17], where the authors identified ranges of parameter values that allow for an organism-level homeostatic steady state in total tumor burden through mutual angio-inhibitory interactions that could yield global dormancy.

Here we propose an alternative model to explain a possible mechanism, whereby secondary tumors might undergo accelerated growth following removal of the primary tumor. Specifically, a possible explanation for the presence of high levels of inhibitors is the normal tissues' response to increased presence of stimulators. This increase would occur to restore the homeostatic ratio of stimulators to inhibitors within the host, returning it to the physiologically normal "off" state. Experimentally, Gonzalez et al. have shown that the ratio of inhibitors to stimulators, such as TSP-1 to VEGF, in serum and in serum minus plasma, is two to three times higher in healthy subjects as it is in cancer patients [18]. Furthermore, some preliminary data suggest that injecting VEGF into mice results in increased levels of angiogenesis inhibitors (Giannoula Klement, personal communication). This suggests that an increase in level of stimulators should be followed by an increase in systemic levels of inhibitors as the body attempts to suppress unnecessary neovascularization.

If the body indeed attempts to compensate for increased stimulator production by increasing production of inhibitors, one can surmise of the following scenario.

Assume the presence in the host of two tumors, a larger primary tumor and a much smaller secondary tumor. Both tumors are promoting to varying degrees the secretion of some stimulators. At some point this combined stimulator output may become sufficient to trigger the body to commence production of inhibitors to restore homeostasis. It is possible that this increase in the systemic level of inhibitors may be sufficient to suppress the smaller tumor.

Surgical removal of the primary tumor would also remove the source of a large amount of stimulators, effectively reducing or even eliminating the need to further produce additional inhibitors to maintain homeostasis. Therefore, the amount of inhibitors would decrease, giving the secondary tumor an opportunity to build up vasculature and grow, perhaps to the point when newly activated systemic production of inhibitors can no longer keep up. At this point, the positive feedback loop of tumor-induced production of self-supporting vasculature would have progressed too far. This mechanism, which is based on normal physiological responses and the body's attempts to restore homeostasis, could provide a possible explanation for why removal of primary tumors sometimes results in accelerated growth of secondary tumors.

This scenario requires the following assumptions to hold:

1. Levels of inhibitors are responsive to the current levels of stimulators, i.e., when the concentration of stimulators increases, the host's body tries to restore a homeostatic ratio of regulators by increasing the levels of inhibitors.
2. There exists a threshold of sensitivity of inhibitors to the presence of angiogenesis stimulators in the body of the host. When the total amount of stimulators is above this threshold, inhibitor secretion increases to compensate for stimulator increase; below this threshold, inhibitor levels return to homeostatic levels [19].

This scenario is summarized in Fig. 5.1.

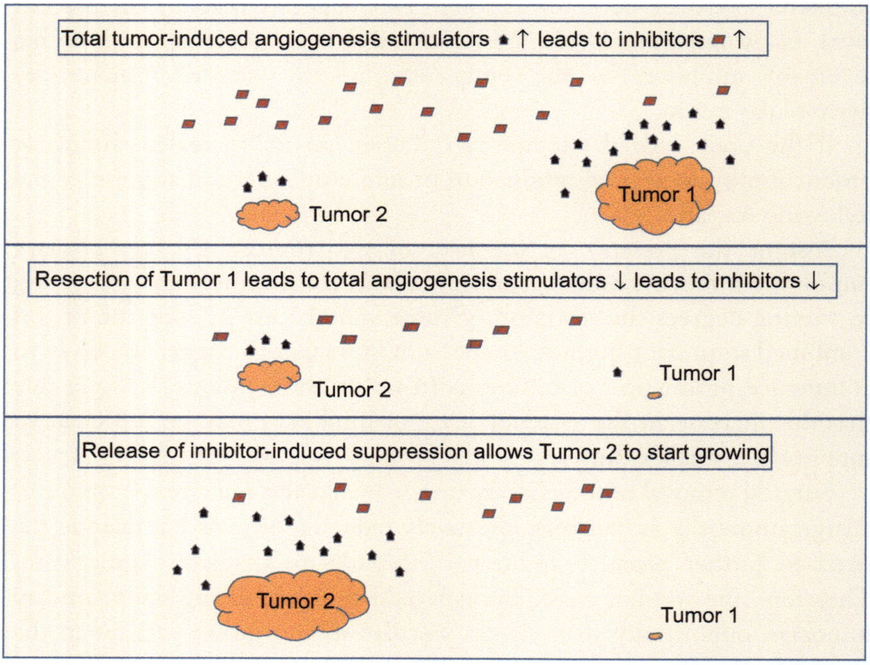

Figure 5.1 Proposed scenario of mechanisms underlying accelerated growth of secondary tumors following resection of a primary tumor. Suppose the host is harboring two tumors. Cumulative amount of angiogenesis regulators could trigger a natural response, where the host would attempt to restore homeostasis of angiogenesis regulators by producing additional inhibitors. Resection of one of the tumors would reduce the total number of angiogenesis stimulators, causing responsive decrease in angiogenesis inhibitors, thus allowing secondary tumor to progress. *(This figure is adapted with permission from Kareva I. Cancer as a disease of homeostasis: an angiogenesis perspective. Ecology and Evolution of Cancer. Elsevier 2017:229–35.)*

This hypothesis is supported by several studies, where the authors evaluated the levels of angiogenesis stimulators and inhibitors before and after surgery. Dipok et al. [21] reported that significant decreases in serum endostatin and basic fibroblast growth factor (bFGF) were observed in postoperative samples of patients with hepatocellular carcinoma, compared to the preoperative values. Similar results were observed by Feldman et al. [22] for patients with colorectal cancer. The authors reported that plasma endostatin levels were significantly higher in the 30 patients compared to controls before surgery. However, none of the patients who remained progression free had elevated endostatin levels at follow-up. Several studies also report that elevated postoperative endostatin levels were associated with poor prognosis for patients with advanced stage nasopharygeal carcinoma [23] and for human malignant gliomas [24], potentially suggesting presence of metastatic tumors.

In this chapter, in order to evaluate this hypothesis, we build a proof-of-concept model, which captures the scenario described above. We identify the necessary assumptions and perform sensitivity analysis to identify key parameters that may be driving system dynamics. We demonstrate that indeed, surgical removal of the primary tumor can lead to a cascade of responses that removes the inhibitory effect on the secondary tumor(s). We conclude with a discussion of therapeutic options.

Model Description

Consider the following model, which takes into account the behavior over time of the following variables: primary tumor $T_1(t)$, its dynamic vasculature $V_1(t)$, tumor-induced angiogenesis stimulators, such as VEGF, described by $S_1(t)$; secondary tumor $T_2(t)$, its dynamic vasculature $V_2(t)$, tumor-induced angiogenesis stimulators $S_2(t)$; and the shared pool of angiogenesis inhibitors, such as PF-4 and TSP-1, described by $I(t)$.

Given our understanding of the system, we expect to see it behave as depicted on Fig. 5.2, where without surgery the overall levels of stimulators remain high, causing high levels of inhibitors and consequent suppression of the secondary tumor. However, in the event of a surgery, the amount of inhibitors decreases, allowing for growth of secondary tumors.

We propose that the change in volume of the primary tumor $T_1(t)$ is described by an Allee-type growth function with a dynamic carrying capacity $\frac{dT_1}{dt} = \lambda_1 T_1(t)\left(\frac{T_1(t)}{m_1} - 1\right)\left(1 - \frac{T_1(t)}{k+V_1(t)}\right)$, where λ_1 is the growth rate of the tumor cells, k is tumor carrying capacity in the absence of

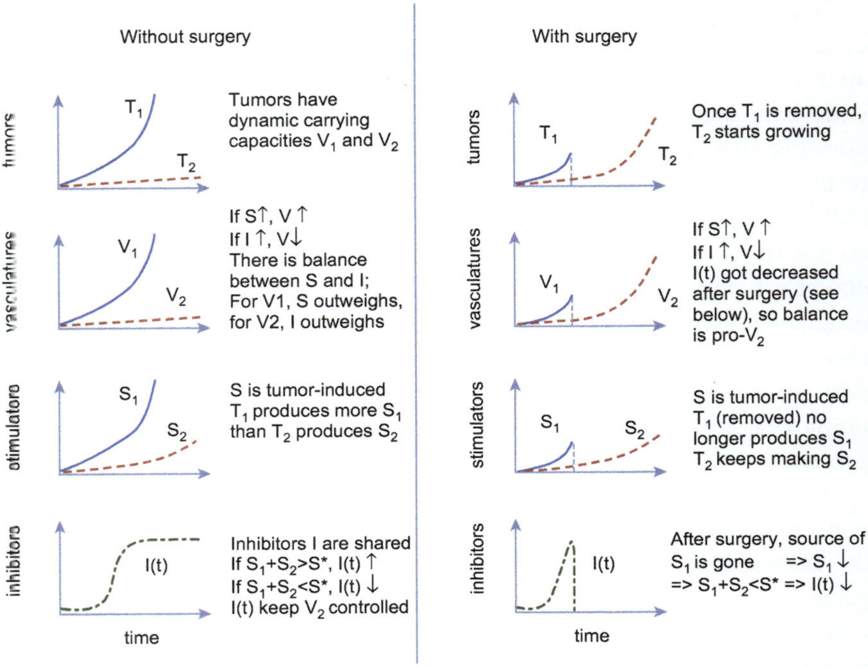

Figure 5.2 Schematic representation of the anticipated dynamics of tumor, neovasculature, stimulators, and inhibitors with and without surgery. We expect that without surgery, the level of inhibitors will remain high, inhibiting secondary tumor growth. However, we predict that after resecting the primary tumor, the level of inhibitors will decrease due to the decrease of overall stimulator levels, causing accelerated growth of the secondary tumor.

neovascularization (the maximum size that a dormant nonvascularized tumor can reach), and $V_1(t)$ is the volume of new vasculature, formed as a result of the activity of angiogenesis stimulators and inhibitors. Parameter $m_1 \neq 0$ describes the threshold of viability of a resected tumor; when $T_1(t) > m_1$, the tumor grows up to the dynamic carrying capacity $V_1(t)$. It reflects an observation that in a three-dimensional culture, depending on the environment and the surrounding signaling, at low concentrations malignant cells can sometimes revert to a nonmalignant phenotype [25–27]. This suggests existence of some threshold of viability, which is accounted for by parameter m_1.

In the previous chapter, we looked at three classes of angiogenesis regulators: stimulators, such as VEGF, blood vessel stabilizers, such as bFGF and PDGF, and angiogenesis inhibitors, such as PF-4 and TSP-1. For the purposes of the question considered here, we can reduce the number of variables without loss of qualitative behavior and combine stimulators and stabilizers into a single class of "stimulators" $S_1(t)$, which act to promote formation of new blood vessels.

Now let us derive equations that describe the dynamics of angiogenesis regulators, starting with stimulators $S(t)$. Here, we describe the dynamics of *tumor-induced* vasculature. That is, we are looking at levels of angiogenesis regulators and corresponding neovasculature created above the body's normal baseline. For this reason, in the absence of a tumor, the level of tumor-induced stimulators will tend to zero. In the presence of a tumor, a tumor-induced inflow of stimulators exists, described by the ratio-dependent functional form $r_1 S_1(t)\frac{T_1(t)/S_1(t)}{1+T_1(t)/S_1(t)} = r_1\frac{S_1(t)T_1(t)}{S_1(t)+T_1(t)}$. This functional form allows tumor-induced production of stimulators to be dependent both on the current amount of stimulators and on the volume of tumor present. We also assume there to be some natural clearance rate of angiogenesis stimulators, described by $\delta S_1(t)$.

The equation for angiogenesis inhibitors $I(t)$ is derived in the following way. Inhibitors $I(t)$ increase if the total sum of angiogenesis stimulators $S_1(t) + S_2(t)$ is greater than some threshold value S^* up to some carrying capacity I_{max}, and decrease to zero if the sum of stimulators $S_1 + S_2$ is less than S^*. As with stimulators, we track only the increases in inhibitors that result from the presence of the tumor(s). This effect is captured by an Allee effect type functional form $r_i I(t)\left(\frac{S_1(t)+S_2(t)}{S^*} - 1\right)\left(1 - \frac{I(t)}{I_{max}}\right)$. Angiogenesis inhibitors are also cleared at some constant rate, described by the term $\delta_3 I(t)$. Notably, as was mentioned above, there exist two classes of mechanisms whereby inhibitors contribute to angiogenesis inhibition, namely, by directly outcompeting angiogenesis stimulators, or by interfering with endothelial cell proliferation and inducing apoptosis. In the previous chapter, we looked at the first mechanism; here, we focus on the second one.

Finally, the equations for the dynamics of tumor-induced vasculature are derived as follows. Increase in vasculature is proportional to the presence of angiogenesis stimulators $\gamma_1 S_1(t)$, and decrease can occur either naturally at some low rate $\xi_1 V_1(t)$ or due to angiogenesis inhibitors, which we propose occurs at a rate $\frac{\nu_1 I(t)}{1+I(t)}$. The balance between stimulators and inhibitors will determine whether tumor-induced vasculature will increase or not. Equations describing the dynamics of the secondary tumor, the corresponding vasculature volume, and the amount of stimulators are derived in the same way, varying only in parameter values.

Noticeably, as was mentioned above, the mechanism of action of inhibitors such as angiostatin and endostatin involves increased apoptosis, which may include some part of the existing neovasculature. This is observed during normal blood vessel formation, when the vasculature is initially overbuilt and then pruned, decreasing in size to a stable level. As one can see from the equations, in the absence of a tumor, $S(t) \rightarrow 0$, and therefore $I(t) \rightarrow 0$, and consequently $V(t) \rightarrow 0$. However, there may be a short transient period, when the value of $V(t)$ briefly becomes negative, indicating excessive inhibitor-induced apoptosis that affects the state of the baseline vasculature. (This also suggests that a secondary tumor could at least transiently suppress the growth of the primary tumor not only through stimulating system-wide production of inhibitors that would outcompete stimulators, but also by temporarily degrading the vasculature of the primary tumor, whose stimulator activity does not compensate quickly enough for the systemic increase in the level of angiogenesis inhibitors.)

Taking into account all of these considerations, we obtain the following system of equations:

$$\underbrace{\frac{dT_1(t)}{dt}}_{\text{tumor growth}} = \underbrace{\lambda_1 T_1(t)\left(\frac{T_1(t)}{m_1} - 1\right)\left(1 - \frac{T_1(t)}{k + V_1(t)}\right)}_{\substack{\text{Allee growth with vasculature dependent}\\ \text{dynamic carrying capacity}}} \qquad \frac{dT_2(t)}{dt} = \lambda_2 T_2(t)\left(\frac{T_2(t)}{m_2} - 1\right)\left(1 - \frac{T_2(t)}{k + V_2(t)}\right)$$

$$\underbrace{\frac{dV_1(t)}{dt}}_{\text{neovasculature}} = \underbrace{\gamma_1 S_1(t)}_{\substack{\text{increase due to}\\ \text{angiogenesis}\\ \text{stimulators}}} \underbrace{- \xi_1 V_1(t)}_{\substack{\text{natural}\\ \text{decay}}} - \underbrace{\frac{\nu_1 I(t)}{1 + I(t)}}_{\substack{\text{inhibitor-induced}\\ \text{vasculature decay}}} \qquad \frac{dV_2(t)}{dt} = \gamma_2 S_2(t) - \xi_2 V_2(t) - \frac{\nu_2 I(t)}{1 + I(t)}$$

$$\underbrace{\frac{dS_1(t)}{dt}}_{\substack{\text{angiogenesis}\\ \text{stimulators}}} = \underbrace{\frac{r_1 S_1(t) T_1(t)}{S_1(t) + T_1(t)}}_{\substack{\text{tumor-induced production}\\ \text{of angiogenesis stimulators}}} - \underbrace{\delta_1 S_1(t)}_{\text{natural decay}} \qquad \frac{dS_2(t)}{dt} = \frac{r_2 S_2(t) T_2(t)}{S_2(t) + T_2(t)} - \delta_2 S_2(t)$$

$$\underbrace{\frac{dI}{dt}}_{\substack{\text{angiogenesis}\\ \text{inhibitors}}} = \underbrace{r_i I(t)\left(\frac{S_1(t) + S_2(t)}{S^*} - 1\right)\left(1 - \frac{I(t)}{I_{\max}}\right)}_{\substack{\text{Allee affect type growth: } I(t) \text{ increases if } S_1(t)+S_2(t)>S^*,\\ \text{up to carrying capacity } I_{max}, \text{ and decreases if } S_1(t)+S_2(t)<S^*.}} - \delta_3 I(t). \tag{5.1}$$

All of the variables and parameters for System (5.1), as well as sample parameter values and initial conditions, are summarized in Table 5.1.

SENSITIVITY ANALYSIS

Just as in previous chapters, we will use sensitivity analysis to evaluate, which parameters in System (5.1) would have the largest impact on the overall system dynamics at various time points. To do so, we again perturb each parameter by a uniformly distributed random variable within the range of 15% of the initial parameter value. The sensitivity indices (fractions of total output variance generated by the uncertainty in the respective parameter value) are again calculated using the Fourier Amplitude Sensitivity Test (FAST) method [28]. Larger values of sensitivity indices indicate larger sensitivity to parameter perturbation at each time point evaluated.

The results of the sensitivity analysis performed on System (5.1) are summarized in Fig. 5.3. Changes over time in the values of the first-order sensitivity indices are plotted in Fig. 5.3A. Owing to large variations in relative sensitivities of different parameters, the results are additionally presented in Fig. 5.3B—D, where sensitivity indices of parameters pertaining to tumor growth are presented in Fig. 5.3B, to vasculature growth in Fig. 5.3C, and to growth of stimulators in Fig. 5.3D. Only parameters with relative sensitivity indices over 0.01% at any time point are reported.

Seven parameters were identified as sensitive to perturbation (Fig. 5.3). Parameters pertaining to tumor growth (Fig. 5.3B) and vasculature growth (Fig. 5.3C) have the largest impact in the beginning of system development, in the time up until $t \approx 150$. However, as the tumors grow, it is parameters pertaining to the dynamics of stimulators that start having the largest impact (Fig. 5.3D), emphasizing the importance of angiogenesis regulators with regards to tumor dynamics.

Out of the seven parameters that were identified using the FAST method (Fig. 5.3A), we will vary only γ_1 and r_1, which describe the rate of stimulator-dependent vasculature formation, and rate of tumor-induced production of stimulators, respectively. In the next section, these two parameters will be the only ones that differ between T_1 and T_2. This will allow us to narrow our investigation down specifically to effects of tumor-induced vasculature formation.

Table 5.1 Variables and parameters for System (5.1)

Parameter/Variable	Description	Sample value	Units
$T_1(t),\ T_2(t)$	Primary and secondary tumor volumes	$T_1(0) = 1$ $T_2(0) = 1$	Volume (i.e., mm^3)
$S_1(t), S_2(t)$	Amount of angiogenesis stimulators produced by respective tumors	$S_1(0) = 0.1$ $S_2(0) = 0.1$	Volume
$V_1(t), V_2(t)$	Tumor(s)-induced vasculature	$V_1(0) = 0.01$ $V_2(0) = 0.01$	Volume
$I(t)$	Angiogenesis inhibitors (TSP-1)	$I(0) = 3$	Volume
λ_1, λ_2	Tumor(s) growth rates	0.04	1/time (i.e., day)
$m_1,\ m_2$	Threshold volume of tumor survival	0.7	Volume
K	Maximum volume of avascular tumor growth	1	Volume
γ_1	Rate of vasculature growth for T_1	2.7	1/time
γ_2	Rate of vasculature growth for T_2	0.5	1/time
ξ_1, ξ_2	Rate of natural vasculature decay	0.09	1/time
ν_1, ν_2	Rate of $I(t)$-induced vasculature inhibition	0.3	1/time
r_1	Rate of tumor-induced production of S_1	0.17	1/time
r_2	Rate of tumor-induced production of S_2	0.11	1/time
δ_1, δ_2	Stimulator (VEGF) clearance rate	0.07	1/time
δ_3	Inhibitor (i.e., PF-4) clearance rate	0.01	1/time
r_i	Rate of increased inhibitor production in response to cumulative $S_1 + S_2$	0.1	1/time
S^*	Inhibitor sensitivity threshold to cumulative $S_1 + S_2$	3	Volume
I_{max}	Carrying capacity for angiogenesis inhibitors	100	Volume

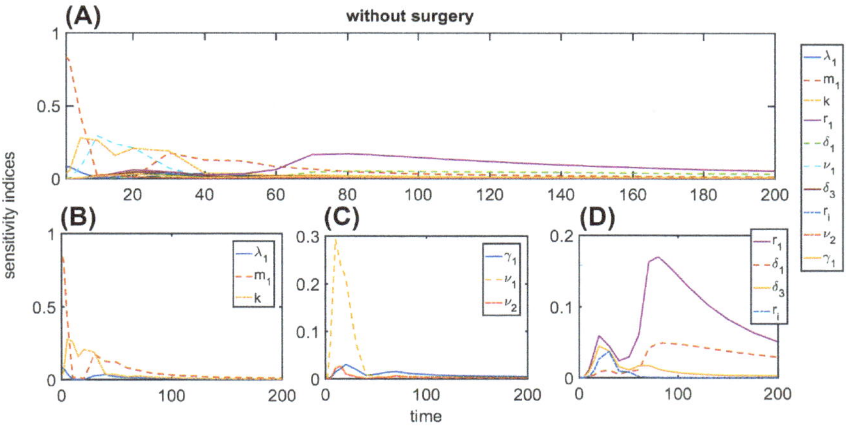

Figure 5.3 First-order sensitivity indices for parameters of System (5.1), obtained using the Fourier Amplitude Sensitivity Test method. Only indices that are over 0.01% are reported. (A) All of the identified parameters that are sensitive to perturbation; (B) sensitive parameters that affect tumor growth; (C) sensitive parameters that affect the growth of neovasculature; (D) sensitive parameters that affect the dynamics of tumor-induced angiogenesis stimulators. All parameters are taken from Table 5.1.

RESULTS

First, we simulated resection of the primary tumor and evaluated, how quickly secondary tumors started growing compared to a scenario with no tumor resection. Tumor resection is simulated by reducing the size of the tumor, its vasculature, and amount of stimulators, by the same percentage at a chosen resection time. These new values are then used as initial conditions to continue with the simulation.

As one can see in Fig. 5.4, without resection, the size of the primary tumor T_1 predictably keeps increasing. Removing 99.99% of the primary tumor at $t = 100$ causes T_1 to decrease below threshold m_1, and the secondary tumor T_2 starts growing sooner than it would have without surgery.

To better understand this effect, let's take a closer look at the dynamics of other components of the system, namely, vasculature, angiogenesis stimulators, and angiogenesis inhibitors, with and without surgery.

In Fig. 5.5, we can see that predictably, the tumor growth dynamics (Fig. 5.5A) corresponds to the tumor's ability to generate neovasculature (Fig. 5.5B). When the amount of stimulators (Fig. 5.5C) increases beyond threshold $S_1 + S_2 > S^*$, inhibitors increase in response until they reach their carrying capacity I_{max}. This increase in inhibitors is not sufficient to

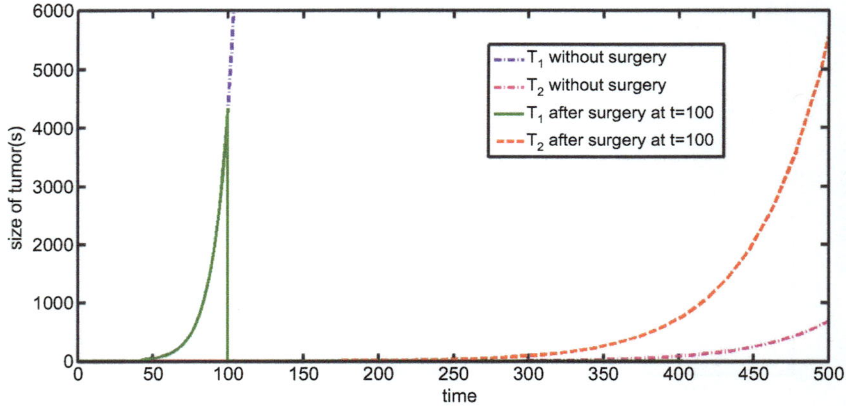

Figure 5.4 Comparison of primary and secondary tumor dynamics without surgery and with surgery at $t = 100$. As one can see, without surgery, the primary tumor keeps growing, but with surgery, secondary tumors start growing sooner.

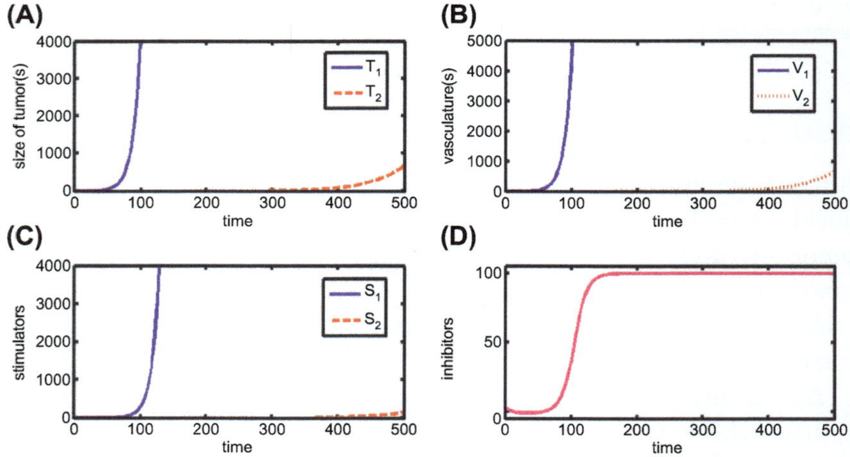

Figure 5.5 Changes over time in tumor size, and corresponding changes in vasculature, angiogenesis stimulators and inhibitors, in a scenario without surgery. (A) Size of primary tumor T_1 and (B) corresponding vasculature continue to increase, since (C) the total amount of stimulators remains high, causing (D) total level of inhibitors to remain elevated, delaying growth of the secondary tumor T_2.

overcome the driving force of stimulators for the primary tumor T_1, but it may be sufficient to temporarily halt the growth of the secondary tumor T_2.

Now let us compare these dynamics to what happens to system components after surgery (Fig. 5.6). As one can see, resection of the primary

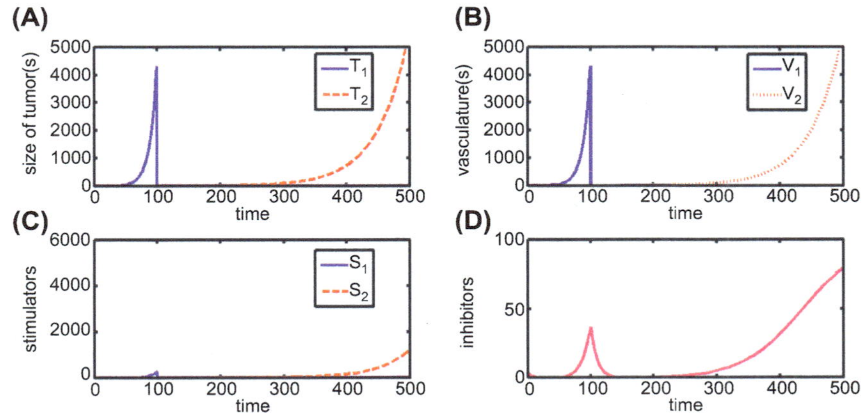

Figure 5.6 Changes over time in tumor sizes, and corresponding changes in vasculature, angiogenesis stimulators and inhibitors, in a scenario with surgery performed to resect the primary tumor at time $t = 100$. (A) Resection of primary tumor T_1 causes reduction in (B) vasculature and (C) amount of stimulators, in turn causing (D) temporary decrease in amount of inhibitors, facilitating growth of the secondary tumor T_2.

tumor T_1 at time $t = 100$ caused a drop in total number of stimulators (Fig. 5.6C), resulting in a consequent drop in angiogenesis inhibitors (Fig. 5.6D). This "window of opportunity" allows the secondary tumor T_2 to produce a sufficient number of stimulators (Fig. 5.6C) to construct neovasculature (Fig. 5.6B) before the total number of stimulators becomes large enough to once again trigger increase in inhibitors (Fig. 5.6D).

It is easy to see from the equation for vasculature dynamics $V'(t)$ why this delay in activation of angiogenesis inhibitors allows building up of sufficient vasculature to overcome the inhibitory effects of $I(t)$. $V'(t) = 0$ when $V^* = \frac{\gamma}{\xi} S(t) - \frac{\nu}{\xi} \frac{I(t)}{1+I(t)}$. As one can see, the dynamics is determined by the balance between $\frac{\gamma}{\xi} S(t)$ and $\frac{\nu}{\xi} \frac{I(t)}{1+I(t)}$, and if the total number of stimulators $S(t)$ becomes very large, it will outweigh the inhibitory effects of $I(t)$, pending values of parameters γ and ν. This explains why even after inhibitors start increasing, they will have little effect if the tumor has had a chance to produce a sufficient number of stimulators.

Time to Surgery and Time to Metastases

Finally, let us look at the effects of timing of surgery on length of time to appearance of secondary tumors. We simulate surgery at $t = 90$, $t = 100$,

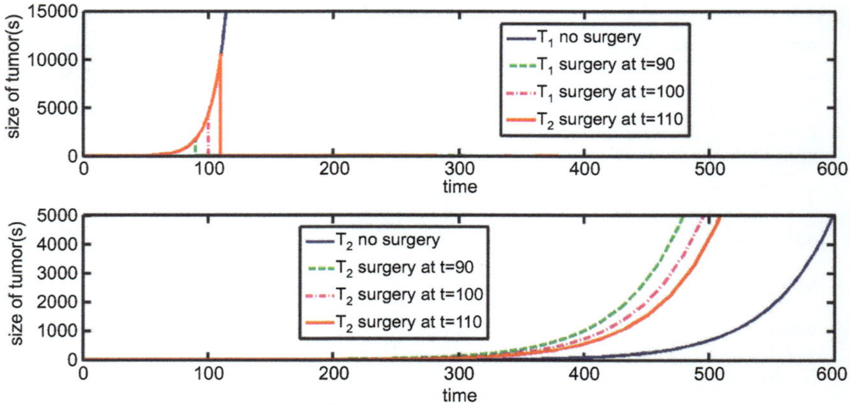

Figure 5.7 Variations in time to appearance of T_2 depending on timing of the surgery. Earlier surgery (top panel) results in faster growing metastases (lower panel). However, later resection of the primary tumor allows it to reach larger sizes, potentially making it unresectable and causing more immediate harm to the patient.

and $t = 110$. As one can see in Fig. 5.7, growth of the secondary tumor is offset by the delay in time of surgery. There exists a trade-off: earlier surgery will reduce the size of the primary tumor T_1 (Fig. 5.7A), but it will also reduce the delay in time to onset of secondary tumor growth (Fig. 5.7B). Conversely, delaying surgery allows the primary tumor to reach a larger size, potentially making it unresectable and more harmful to the patient. Within this construct, it is important to note that resecting a primary tumor does not induce growth of secondary tumors but simply reduces time to onset of metastatic growth.

To further understand how the system changes after surgery compared to before surgery, we perform additional sensitivity analysis to identify which parameters become more or less important. Results are reported in Fig. 5.8.

Sensitivity indices are predictably identical in the time before surgery, so we report only the changes that occur after surgery. As one can see in Fig. 5.8, parameters r_1 and r_i have greater impact on system dynamics after the surgery than in the control case. This indicates the high importance of both the rate at which the tumor produces stimulators, and the ability of inhibitors to respond in time. The rate at which the host can adapt to the change in stimulators becomes a crucial factor in determining whether secondary tumors will experience accelerated growth in the time after surgery.

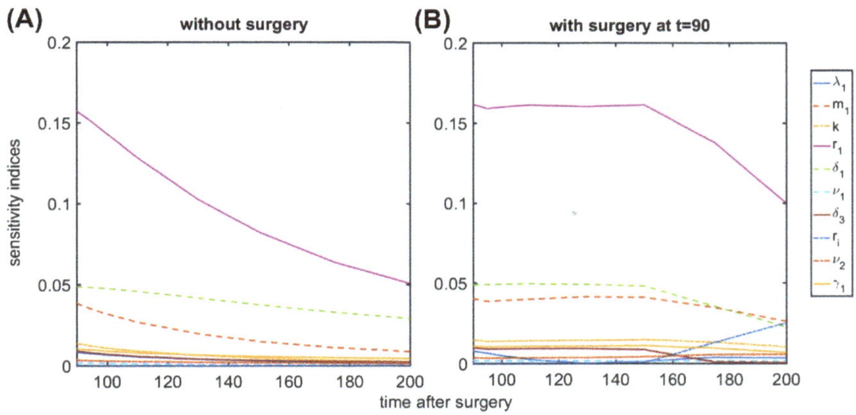

Figure 5.8 Comparison of first-order sensitivity indices for parameters of System (5.1) obtained using the Fourier Amplitude Sensitivity Test (FAST) method for (A) simulation of system dynamics without intervention and (B) simulation of system dynamics with surgery at $t = 90$.

THERAPEUTIC IMPLICATIONS

Even though there may exist a trade-off between surgery timing and accelerated growth of metastatic tumors, in the presence of a clinically apparent tumor the question is not whether to resect it or wait, but how to address the danger of potential metastases. In this system, accelerated growth of secondary tumors following the resection of a primary tumor occurs when the total number of angiogenesis stimulators decreases so much as to cause decrease in the number of inhibitors. One possible way to address this issue would be to directly target the source of stimulators after surgery. Targeting the stroma with maintenance therapy, such as for ALL, where a period of high-intensity induction is followed by 2–3 years of lower-dose, higher-frequency therapy [29] or metronomic therapy, which similarly involves administering lower doses of chemotherapeutic agents at more frequent time intervals [30–32], after surgical resection might halt the growth of secondary tumors and reduce the risk of metastatic recurrences.

This will be the subject of the final chapter.

REFERENCES

[1] Bedenne L, Michel P, Bouché O, Milan C, Mariette C, Conroy T, et al. Chemoradiation followed by surgery compared with chemoradiation alone in squamous cancer of the esophagus: FFCD 9102. Journal of Clinical Oncology. American Society of Clinical Oncology 2007;25(10):1160–8.

[2] Peeters CF, de Waal RM, Wobbes T, Ruers TJ. Metastatic dormancy imposed by the primary tumor: does it exist in humans? Annals of surgical oncology. Springer 2008;15(11):3308–15.
[3] Retsky M, Demicheli R, Hrushesky W, Baum M, Gukas I. Surgery triggers outgrowth of latent distant disease in breast cancer: an inconvenient truth? Cancers. Molecular Diversity Preservation International 2010;2(2):305–37.
[4] Peeters CF, De Waal RM, Wobbes T, Westphal JR, Ruers TJ. Outgrowth of human liver metastases after resection of the primary colorectal tumor: a shift in the balance between apoptosis and proliferation. International Journal of Cancer. Wiley Online Library 2006;119(6):1249–53.
[5] Van der Bij GJ, Oosterling SJ, Beelen RH, Meijer S, Coffey JC, van Egmond M. The perioperative period is an underutilized window of therapeutic opportunity in patients with colorectal cancer. Annals of Surgery. LWW 2009;249(5):727–34.
[6] Demicheli R, Fornili M, Ambrogi F, Higgins K, Boyd JA, Biganzoli E, et al. Recurrence dynamics for non–small-cell lung cancer: effect of surgery on the development of metastases. Journal of Thoracic Oncology. Elsevier 2012;7(4):723–30.
[7] Chiarella P, Bruzzo J, Meiss RP, Ruggiero RA. Concomitant tumor resistance. Cancer Letters. Elsevier 2012;324(2):133–41.
[8] Ruggiero RA, Bruzzo J, Chiarella P, Bustuoabad OD, Meiss RP, Pasqualini CD. Concomitant tumor resistance: the role of tyrosine isomers in the mechanisms of metastases control. Cancer Research. AACR 2012;72(5):1043–50.
[9] Prehn RT. Two competing influences that may explain concomitant tumor resistance. Cancer Research. AACR 1993;53(14):3266–9.
[10] Kareva I. Primary and metastatic tumor dormancy as a result of population heterogeneity. Biology Direct. BioMed Central 2016;11(1):37.
[11] Peterson JE, Zurakowski D, Italiano Jr JE, Michel LV, Connors S, Oenick M, et al. VEGF, PF4 and PDGF are elevated in platelets of colorectal cancer patients. Angiogenesis. Springer 2012;15(2):265–73.
[12] Bremnes RM, Camps C, Sirera R. Angiogenesis in non-small cell lung cancer: the prognostic impact of neoangiogenesis and the cytokines VEGF and bFGF in tumours and blood. Lung Cancer. Elsevier 2006;51(2):143–58.
[13] Carmeliet P. VEGF as a key mediator of angiogenesis in cancer. Oncology. Karger Publishers 2005;69(Suppl. 3):4–10.
[14] Folkman J. Angiogenesis inhibitors generated by tumors. Molecular medicine. The Feinstein Institute for Medical Research 1995;1(2):120.
[15] Chen C, Parangi S, Tolentino MJ, Folkman J. A strategy to discover circulating angiogenesis inhibitors generated by human tumors. Cancer Research. AACR 1995;55(19):4230–3.
[16] Folkman J. Angiogenesis: an organizing principle for drug discovery? Nature Reviews Drug Discovery. Nature Publishing Group 2007;6(4):273–86.
[17] Benzekry S, Gandolfi A, Hahnfeldt P. Global dormancy of metastases due to systemic inhibition of angiogenesis. PLoS One. Public Library of Science 2014;9(1):e84249.
[18] Gonzalez F, Rueda A, Sevilla I, Alonso L, Villarreal V, Torres E, et al. Shift in the balance between circulating thrombospondin-1 and vascular endothelial growth factor in cancer patients: relationship to platelet alpha-granule content and primary activation. The International Journal of Biological Markers 2003;19(3):221–8.
[19] Peterson JE, Zurakowski D, Italiano JE, Michel LV, Fox L, Klement GL, et al. Normal ranges of angiogenesis regulatory proteins in human platelets. American Journal of Hematology. Wiley Online Library 2010;85(7):487–93.
[20] Kareva I. Cancer as a disease of homeostasis: an angiogenesis perspective. Ecology and Evolution of Cancer. Elsevier 2017:229–35.

[21] Dhar DK, Ono T, Yamanoi A, Soda Y, Yamaguchi E, Rahman MA, et al. Serum endostatin predicts tumor vascularity in hepatocellular carcinoma. Cancer. Wiley Online Library 2002;95(10):2188–95.
[22] Feldman AL, Alexander Jr HR, Bartlett DL, Kranda KC, Miller MS, Costouros NG, et al. A prospective analysis of plasma endostatin levels in colorectal cancer patients with liver metastases. Annals of Surgical Oncology. Springer 2001;8(9):741–5.
[23] Mo H-Y, Luo D-H, Qiu H-Z, Liu H, Chen Q-Y, Tang L-Q, et al. Elevated serum endostatin levels are associated with poor survival in patients with advanced-stage nasopharyngeal carcinoma. Clinical Oncology. Elsevier 2013;25(5):308–17.
[24] Morimoto T, Aoyagi M, Tamaki M, Yoshino Y, Hori H, Duan L, et al. Increased levels of tissue endostatin in human malignant gliomas. Clinical Cancer Research. AACR 2002;8(9):2933–8.
[25] Bissell MJ, Radisky DC, Rizki A, Weaver VM, Petersen OW. The organizing principle: microenvironmental influences in the normal and malignant breast. Differentiation. Elsevier 2002;70(9–10):537–46.
[26] Weaver VM, Petersen OW, Wang F, Larabell C, Briand P, Damsky C, et al. Reversion of the malignant phenotype of human breast cells in three-dimensional culture and in vivo by integrin blocking antibodies. The Journal of Cell Biology. Rockefeller University Press 1997;137(1):231–45.
[27] Lee GY, Kenny PA, Lee EH, Bissell MJ. Three-dimensional culture models of normal and malignant breast epithelial cells. Nature Methods. Nature Publishing Group 2007;4(4):359–65.
[28] Cukier R, Fortuin C, Shuler KE, Petschek A, Schaibly J. Study of the sensitivity of coupled reaction systems to uncertainties in rate coefficients. I Theory. The Journal of Chemical Physics. AIP Publishing 1973;59(8):3873–8.
[29] Toyoda Y, Manabe A, Tsuchida M, Hanada R, Ikuta K, Okimoto Y, et al. Six months of maintenance chemotherapy after intensified treatment for acute lymphoblastic leukemia of childhood. Journal of Clinical Oncology. American Society of Clinical Oncology 2000;18(7):1508–16.
[30] Kareva I, Waxman DJ, Klement GL. Metronomic chemotherapy: an attractive alternative to maximum tolerated dose therapy that can activate anti-tumor immunity and minimize therapeutic resistance. Cancer Letters. Elsevier 2015;358(2):100–6.
[31] Pasquier E, Kavallaris M, André N. Metronomic chemotherapy: new rationale for new directions. Nature Reviews Clinical Oncology. Nature Publishing Group 2010;7(8):455–65.
[32] Klement GL. Eco-evolution of cancer resistance. Science translational medicine. American Association for the Advancement of Science 2016;8(327):327fs5.

CHAPTER 6

Cancer as a Systemic Disease That Requires a Systemic Approach

Contents

ORIGINS OF MAXIMUM TOLERATED DOSE PROTOCOL FOR CHEMOTHERAPY ADMINISTRATION

Cytotoxic chemotherapy acts by interfering with cell growth or inducing cell death of dividing cells, often through targeting different stages of the cell cycle. Standard protocol for cancer treatment is maximum tolerated dose (MTD). It involves administering the highest dose of the chemotherapeutic agent that can be tolerated by the patient and requires prolonged intervals between treatments to allow normal tissues to recover and repair themselves.

This approach was derived from success in treating acute lymphoblastic leukemia (ALL) in children, a rare case when the cancer clone can be

Understanding Cancer From a Systems Biology Point of View
ISBN 978-0-12-813673-7
http://dx.doi.org/10.1016/B978-0-12-813673-7.00006-9

completely eradicated [1]. It was first established in 1960s, before there was a realization that

1. most tumors are genetically heterogeneous (this realization came with the explosion of the field of molecular biology in the 1970s)
2. most complex and difficult to treat tumors engage and modify their environment.

Therefore, MTD works best for more genetically homogeneous tumors (i.e., ALL, testicular cancer, Hodgkin disease, B cell non-Hodgkin lymphomas) but not for complex cancers, such as sarcomas, breast, pancreatic, and lung cancer.

WHAT DO WE NOW KNOW ABOUT MOST SOLID TUMORS?

Firstly, tumors that are most difficult to eradicate are highly genetically heterogeneous, and they keep mutating over time [2—4]. Therefore, administration of traditional chemotherapy results in killing therapy-sensitive clones, leaving behind therapy-resistant population—in a way, MTD is the best way to select for therapy-resistant cancer cells.

Furthermore, transformed malignant cancer cells do not make up the entire tumor. In fact, a large portion of the tumor is transformed stromal cells, which support tumor growth. They engage and modify their microenvironment through changing the pH, transforming fibroblasts to induce production of angiogenesis stimulators, effectively creating a niche for themselves within the tissue [5—8].

Therefore, cancer can be thought of as an evolving ecological system, where a population of genetically heterogeneous cells

- interact with each other and their environment
- compete for nutrients and space
- interact with predators (the immune system)
- migrate (metastases)

all are within the ecosystem of the human body. From this point of view, when a cancer kills its host, it *commits evolutionary suicide*, a situation when short-term adaptations lead to eventual destruction of the habitat, on which the cancer depends.

In the previous chapters, we looked at a particular aspect of cancer ecology, namely, the tumor's interactions with its natural predator, the immune system, and mechanisms of immune evasion, when the prey (cancer) outcompetes the predator (immune cells) for shared resources (glucose). We saw that this mechanism can be quite effective in providing the tumor with a way to avoid immune predation and continue unrestrained growth.

CANCER AS AN ECOSYSTEM: LESSONS FROM CONSERVATION BIOLOGY

If cancer is an ecosystem, we can apply insights from ecological literature on how to approach it. Specifically, conservation biology tells us that the best way to kill a species is not to target individuals directly but to *target their environment*. In the case of cancer, our goal is to extinguish these species from the ecosystem of the human body, and the task becomes to find ways to modify the environment in such a way as to undermine support for cancer cells without killing the host.

This perspective therefore requires reformulating treatment goals: it is not short-term tumor burden reduction that we are after, but *long-term patient survival*—and the two are not the same.

METRONOMIC (MAINTENANCE) CHEMOTHERAPY AS A WAY TO TARGET TUMOR MICROENVIRONMENT

All the cancer cells depend on supporting cells of its microenvironment to provide signaling for blood vessel formation, ensuring a supply of food and oxygen. This makes the microenvironment itself an attractive therapeutic target. One approach that allows doing this involves administration of same chemotherapeutic drugs used in MTD but on a different schedule, giving the patients lower doses at more frequent time intervals. This approach is called "metronomic" chemotherapy, and it has numerous advantages. It is much easier on patients, improving patient quality of life; it allows repurposing of drugs, making it cheaper, and most importantly, it has important therapeutic benefits compared to the standard MTD approach. These include decreased angiogenesis, preservation and augmentation of antitumor immunity, and potential for decreased therapeutic resistance.

DECREASED ANGIOGENESIS

Therapeutically resistant tumors engage their microenvironment by recruiting normal tissue cells, such as fibroblasts, pericytes, histiocytes, platelets, and hematopoietic progenitor cells. Through engagement and modification of their microenvironment, cancer cells "simulate" conditions of a wound [11], evoking normal physiological responses such as new blood vessel formation without allowing for angiogenesis termination.

As was mentioned in the previous chapters, tumors cannot grow beyond a size of 1−2 mm^3 without securing their own blood supply. The process of both normal and pathological vascularization is mediated by tightly regulated sequential release of pro- and antiangiogenesis growth factors from platelets, which need to bind to a shared pool of binding sites to signal. Many tumors induce increased production of angiogenesis stimulators, which may outcompete inhibitors for binding sites in the tumor microenvironment, much like in a game of musical chairs. This prevents angiogenesis termination, allowing the tumor to keep growing.

Low-dose high-frequency chemotherapy aims to target not the cancer cells themselves but the supporting stroma, which is the source of angiogenesis regulators. Removing the source of stimulators gives inhibitors an opportunity to signal and terminate the process of blood vessel formation.

IMPROVED ANTITUMOR IMMUNITY

The functionality of the immune system can be compromised by high-dose chemotherapy as immune cells can be ablated by the cytotoxic drugs, preventing them from pursuing therapy-resistant cancer cells. However, metronomically administered chemotherapy can target and thus reduce the numbers of immunosuppressive regulatory T cells (Tregs). It can also promote maturation of antigen-presenting cells, and most importantly, improve activation and functionality of key cytotoxic immune cells, particularly the natural killer (NK) cells of the innate immune response, and CD8+T cells of the adaptive arm of the immune system (a more detailed description of the mechanisms of activation of antitumor immune responses was provided in an earlier chapter).

For instance, Doloff and Waxman [9] demonstrated that administration of metronomic cyclophosphamide every 6 days (Q6day cycle) resulted in significant recruitment and activation of natural killer (NK) cells, dendritic cells, and macrophages, a response that was accompanied by dramatic regression of implanted glioma xenografts. More frequent administration of cyclophosphamide inflicted severe damage on NK cells themselves, while Q9day and Q12day schedule eventually resulted in tumor escape. Similar results were obtained for increasing activation and functionality of CD8+T cells [10].

However, finding the right timing is only part of the task—the authors also showed that efficacy of metronomic chemotherapy for activation of the antitumor immune response was both *time- and dose-dependent*. A dose that was too low caused insufficient immune stimulation, while a dose that was too high increased immune cell mortality. Therefore, appropriate timing and dosing schedule might dramatically improve disease outcome by both engaging and protecting antitumor immune responses. Unfortunately, finding this sweet spot for different cancer types still remains a challenge.

Furthermore, as was discussed earlier in this book, even the well-activated cytotoxic lymphocytes can become incapable of performing cytotoxic function if they are outcompeted for nutrients in the tumor microenvironment. High-frequency low-dose chemotherapy can further augment antitumor immune responses by removing some competition for nutrients, thus facilitating tumor infiltration and improving efficacy of the immune response.

DECREASED THERAPEUTIC RESISTANCE

Therapeutic resistance remains a major obstacle in treating cancers, particularly in advanced stages. MTD applies severe selective pressure to the heterogeneous tumor cell population, thus effectively killing sensitive cells and leaving therapy-resistant clones behind; typically the dose that can eliminate all the tumor cells would be so toxic it might kill the patient as well. Metronomic chemotherapy targets stroma as opposed to individual cells, thus inflicting equal damage on both sensitive and resistant clones ("cutting the branch" all the cancer cells are sitting on), decreasing the chance for development of therapeutic resistance. This mechanism, combined with therapy-induced augmented activation and improved activity of cytotoxic lymphocytes, may provide a way to not only decrease emergence of therapeutic resistance but also allow the immune system to infiltrate the tumor and target therapy-resistant cells from within.

CLINICAL TRIALS

A number of Phase I and II clinical trials for metronomic administration of chemotherapy are ongoing or completed on www.clinicaltrials.gov. One of the most notable completed randomized controlled Phase III trials, CAIRO3, was conducted by the Dutch Colorectal Cancer Group and involved maintenance treatment with capecitabine and bevacizumab in metastatic colorectal cancer [12]. In the trial, patients were randomly assigned to either maintenance treatment with capecitabine and bevacizumab (maintenance group) or observation (observation group). According to the published report, "The primary endpoint of median PFS2 was significantly improved in patients on maintenance treatment, and was 8·5 months in the observation group and 11·7 months in the maintenance group (HR 0·67, 95% CI 0·56–0·81, $P < 0{\cdot}0001$). This difference remained significant when any treatment after PFS1 was considered. Maintenance treatment was well tolerated... The global quality of life did not deteriorate during maintenance treatment and was clinically not different between treatment groups."

POTENTIAL SYNERGY BETWEEN IMMUNOTHERAPY AND CHEMOTHERAPY

It is likely that engaging the body's antitumor immunity might hold the key to targeting therapy-resistant cells, provided that cytotoxic cells are not ablated by the treatment and are capable of tumor infiltration. A possible way to achieve this effect can be to combine metronomically administered chemotherapy with immunotherapy, and in particular, with one of the most promising types of immunotherapy, the immune checkpoint inhibitors. Immune checkpoint inhibitors target the mechanisms that regulate immune cell activation and cytotoxic function against self-antigens as a protection against autoimmune disease. Reducing some degree of checkpoint activity has been shown to significantly augment immune responses [13], leading to improved outcomes in cancer patients [14]. Some of the most well-researched targets involve cytotoxic T-lymphocyte antigen-4 (CTLA-4) cell surface receptor, the programmed cell death-1 (PD-1) surface receptor, or the corresponding soluble PD-L1 or PD-L2 ligands [15].

CYTOTOXIC T-LYMPHOCYTE ANTIGEN-4 INHIBITION

Activation of naïve T cells is mediated by interactions with antigen-presenting cells (APCs), such as dendritic cells, which result in formation of specialized structure at points of contact between T cells and APCs (immunological synapses) [16]. T cells express proteins on the cell surface that provide costimulatory signals to activate or suppress T cell activation and survival, including cluster of differentiation 28 (CD28) and CTLA-4. Binding of CD80 or CD86 receptors on the APC surface to CTLA-4 on T cells results in blocking T cell activation. Alternatively, binding of CD80/86 to T cell receptor CD28 results in increasing T cell activation [15,17,18]. Pharmacological blockade of CTLA-4 gives competitive advantage to CD28, resulting in increased T cell activity. This mechanism is summarized in Fig. 6.1.

CTLA-4 immune checkpoint inhibitor interferes with potentially autoreactive T cells in the earlier stages of T cell activation, primarily in lymph nodes [18,19]. The first FDA-approved checkpoint inhibitor, which acts by blocking the CTLA-4 receptor, is ipilimumab, a monoclonal antibody approved in 2011 for treatment of unresectable or metastatic melanoma [20]. It has shown much promise and is currently undergoing clinical trials for treatment of other cancers, including lung, kidney, and prostate cancers (Phase III), as well as cervical, colorectal, gastric, pancreatic, ovarian, and urothelial cancers (Phase II).

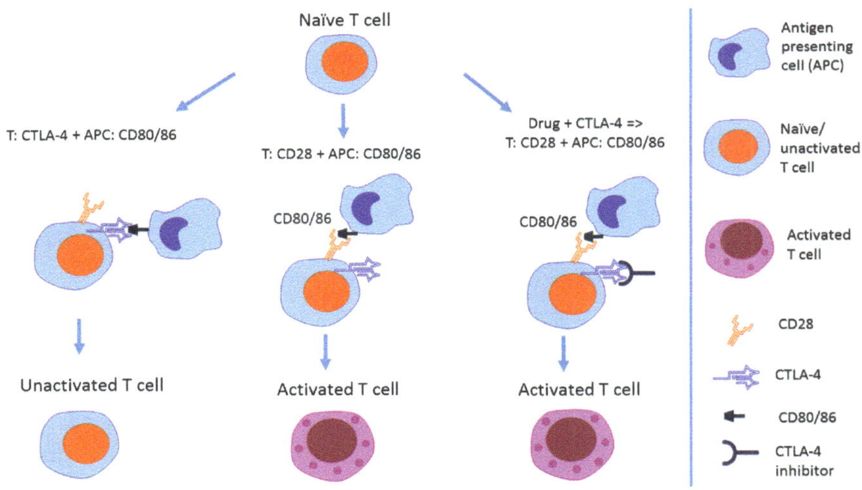

Figure 6.1 Mechanism of action of cytotoxic T-lymphocyte antigen-4 (CTLA-4) checkpoint inhibitor. CTLA-4 and cluster of differentiation 28 (CD28) are receptors expressed on T cell surface; CD80/86 are receptors on the surface of antigen-presenting cells (APCs). When CTLA-4 comes in contact with CD80/86 receptors, the T cell remains unactivated. Interaction of CD28 on T cell surface with CD80/86 on the APC cell surface results in T cell activation. Pharmacological blockade of CTLA-4 on T cell surface increases likelihood of CD28–CD80/86 binding, resulting in maintenance of T cell activation.

PD-1, PD-L1, AND PD-L2 INHIBITION

Programmed cell death protein 1 (PD-1) is a cell surface receptor that regulates T cell activation through binding to soluble ligands PD-L1 and PD-L2 [15,21]. Similarly to CTLA-4, PD-1 signaling interferes with T cell proliferation, glucose metabolism, and cytokine signaling. It promotes apoptosis in antigen-specific T cells and reduces apoptosis of Tregs, indirectly increasing immune cell regulation [15,22–24]. Increased expression of PD-1 is one of the hallmarks of T cell exhaustion [25]. Some important differences between the two mechanisms lie in the fact that while CTLA-4 regulates T cells activation early in the immune response, primarily in lymph nodes, PD-1 affects T cell effector responses in the peripheral tissues [19]. Furthermore, while CTLA-4 expression is restricted to T cells, PD-1 is expressed by T cells, B cells, and macrophages; it can also be expressed on other cells, including some tumor cells [21,26,27].

The PD-L1 and PD-L2 ligands are expressed on leukocytes, macrophages as dendritic cells; PD-L1 has also been found on many tumor types and has been associated with poor prognosis [27]. Since PD-1 ligands are expressed in peripheral tissues, these complexes are believed to maintain self-tolerance in locally infiltrated tissues [19]. Pharmacological interventions for augmenting immune cell responses involve either blocking the PD-1 cell surface receptor or blocking the soluble ligands PD-L1 and PD-L2 to interfere with complex formation. The currently approved anti-PD-1 drugs include pembrolizumab for treatment of unresectable or metastatic melanoma and nivolumab for second and third line treatment of unresectable or metastatic melanoma, or for treatment of metastatic non–small cell lung cancer. Anti-PD-L1 drugs durvalumab and atezolizumab have entered phase III clinical trials for head and neck cancer (durvalumab), bladder cancer (atezolizumab), and lung cancer (both), and Phase II trials for colorectal cancer and glioblastoma (durvalumab) and kidney cancer (atezolizumab). Avelumab was recently approved by FDA for treatment of metastatic Merkel cell carcinoma and bladder cancer.

Immune checkpoint inhibition has been gaining success, particularly in treatment of metastatic cancers, which still remain the cause of majority of cancer-related deaths. It is possible that the therapeutic effect can be further augmented through combination with metronomically administered chemotherapy due to synergistic mechanisms outlined in the following section.

COMBINATION: METRONOMIC CHEMOTHERAPY AND CHECKPOINT INHIBITORS

Based on the mechanisms of action of both metronomic chemotherapy and immune checkpoint inhibitors, one can surmise that a combination of the two therapeutic approaches could have a promising synergistic effect for the following reasons:

1. Both immune checkpoint inhibitors and metronomic chemotherapy increase immune cell activation. While metronomic chemotherapy can promote tumor-specific immune activation, concurrent administration of immune checkpoint inhibitors would maintain the activated state of T cells.
2. Administration of metronomic chemotherapy would allow reducing competition for nutrients between tumor and immune cells via gradual elimination of tumor cells (removing competitors), thus facilitating tumor infiltration by the immune cells.
3. Experimental evidence [28] has shown that blocking PD-L1 directly on tumors dampens glycolysis, further giving cytotoxic lymphocytes a competitive advantage.

Based on these arguments, it is likely that administration of a combination of immune checkpoint inhibitors and chemotherapy on a metronomic schedule would create a synergistic effect that could augment and surpass the therapeutic effect of either therapy alone. Specifically, this approach may provide an avenue for targeting therapy-resistant cells without unacceptable toxicity, resulting in higher treatment compliance, improved long-term outcomes for difficult to treat cancers, and improved patient quality of life (Fig. 6.2).

THINKING OF CANCER AS A SYSTEMIC DISEASE

Hopefully, this book has highlighted that the processes that make cancer such a difficult disease are rooted in its systemic nature. Cancer engages and modifies various components of the human body, which interact and influence each other. Therefore, targeting any one aspect of the system will cause changes throughout its entirety. This balancing act is a challenge that needs to be carefully considered in finding appropriate treatment strategies. Incorporating this kind of systemic thinking will provide us with approaches and solutions that would not have been possible otherwise.

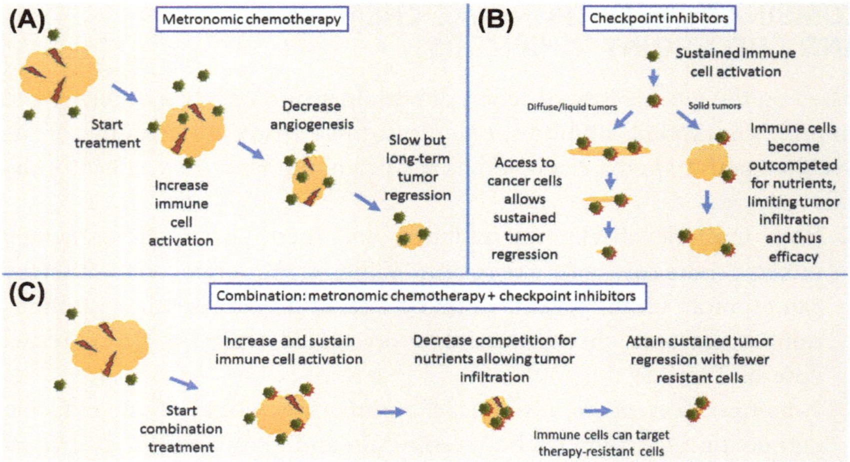

Figure 6.2 Proposed synergistic effects of metronomic chemotherapy combined with checkpoint inhibition. (A) Metronomic chemotherapy causes decrease in angiogenesis and augments antitumor immunity, resulting in slow but long-term tumor regression. (B) Checkpoint inhibition results in sustained immune cell activation; the therapeutic effect is still limited in solid tumors due to low tumor infiltration. (C) Combination of the two therapeutic modalities could create a synergistic effect: metronomic chemotherapy would increase immune activation and facilitate tumor infiltration by removing metabolic competition between tumor and immune cells. Checkpoint inhibition would maintain immune activation, allowing effective elimination of therapy-resistant cells by immune cells that are now capable of infiltrating tumor core.

REFERENCES

[1] Skipper HE, Schabel F, Mellett LB, Montgomery JA, Wilkoff LJ, Lloyd HH, et al. Implications of biochemical, cytokinetic, pharmacologic, and toxicologic relationships in the design of optimal therapeutic schedules. Cancer Chemotherapy Reports Part 1 1970;54(6):431–50.

[2] Marusyk A, Polyak K. Tumor heterogeneity: causes and consequences. Biochimica et Biophysica Acta (BBA)-Reviews on Cancer. Elsevier 2010;1805(1):105–17.

[3] Klein CA, Blankenstein TJ, Schmidt-Kittler O, Petronio M, Polzer B, Stoecklein NH, et al. Genetic heterogeneity of single disseminated tumour cells in minimal residual cancer. The Lancet. Elsevier 2002;360(9334):683–9.

[4] Burrell RA, McGranahan N, Bartek J, Swanton C. The causes and consequences of genetic heterogeneity in cancer evolution. Nature. Nature Publishing Group 2013;501(7467):338–45.

[5] Kareva I. What can ecology teach us about cancer? Translational Oncology 2011;4(5).

[6] Kareva I. Cancer ecology: niche construction, keystone species, ecological succession, and ergodic theory. Biological Theory. Springer 2015;10(4):283–8.

[7] Kareva I, Waxman DJ, Klement GL. Metronomic chemotherapy: an attractive alternative to maximum tolerated dose therapy that can activate anti-tumor immunity and minimize therapeutic resistance. Cancer Letters. Elsevier 2015;358(2):100–6.

[8] Klement GL. Eco-evolution of cancer resistance. Science translational medicine. American Association for the Advancement of Science 2016;8(327):327fs5.

[9] Doloff JC, Waxman DJ. Vegf receptor inhibitors block the ability of metronomically dosed cyclophosphamide to activate innate immunity−induced tumor regression. Cancer Research AACR 2012;72(5):1103−15.

[10] Wu J, Waxman DJ. Metronomic cyclophosphamide eradicates large implanted GL261 gliomas by activating antitumor Cd8+ T-cell responses and immune memory. Oncoimmunology. Taylor & Francis 2015;4(4):e1005521.

[11] Dvorak HF. Tumors: wounds that do not heal. New England Journal of Medicine. Mass Medical Society 1986;315(26):1650−9.

[12] Simkens LH, van Tinteren H, May A, ten Tije AJ, Creemers G-J, Loosveld OJ, et al. Maintenance treatment with capecitabine and bevacizumab in metastatic colorectal cancer (CAIRO3): a phase 3 randomised controlled trial of the Dutch Colorectal Cancer Group. The Lancet. Elsevier 2015;385(9980):1843−52.

[13] Pardoll DM. The blockade of immune checkpoints in cancer immunotherapy. Nature Reviews Cancer. Nature Publishing Group 2012;12(4):252−64.

[14] Eggermont AM, Maio M, Robert C. Immune checkpoint inhibitors in melanoma provide the cornerstones for curative therapies. Seminars in Oncology. Elsevier 2015:429−35.

[15] Buchbinder EI, Desai A. CTLA-4 and PD-1 pathways: similarities, differences, and implications of their inhibition. American Journal of Clinical Oncology. LWW 2016;39(1):98−106.

[16] Murphy K, Weaver C. Janeway's immunobiology. Garland Science 2016.

[17] Lim TS, Goh JKH, Mortellaro A, Lim CT, Hämmerling GJ, Ricciardi-Castagnoli P. CD80 and CD86 differentially regulate mechanical interactions of T-cells with antigen-presenting dendritic cells and B-cells. PLoS One. Public Library of Science 2012;7(9):e45185.

[18] Krummel MF, Allison JP. CD28 and CTLA-4 have opposing effects on the response of T cells to stimulation. The Journal of Experimental Medicine. Rockefeller University Press 1995;182(2):459−65.

[19] Fife BT, Bluestone JA. Control of peripheral T-cell tolerance and autoimmunity via the CTLA-4 and PD-1 pathways. Immunological Reviews. Wiley Online Library 2008;224(1):166−82.

[20] Cameron F, Whiteside G, Perry C. Ipilimumab. Drugs. Springer 2011;71(8):1093−104.

[21] Fife BT, Pauken KE, Eagar TN, Obu T, Wu J, Tang Q, et al. Interactions between PD-1 and PD-L1 promote tolerance by blocking the TCR−induced stop signal. Nature Immunology. Nature Publishing Group 2009;10(11):1185−92.

[22] Takahashi T, Tagami T, Yamazaki S, Uede T, Shimizu J, Sakaguchi N, et al. Immunologic self-tolerance maintained by CD25+ CD4+ regulatory T cells constitutively expressing cytotoxic T lymphocyte−associated antigen 4. The Journal of Experimental Medicine. Rockefeller University Press 2000;192(2):303−10.

[23] Huber S, Hoffmann R, Muskens F, Voehringer D. Alternatively activated macrophages inhibit T-cell proliferation by Stat6-dependent expression of PD-L2. Blood. Am Soc Hematology 2010;116(17):3311−20.

[24] Parry RV, Chemnitz JM, Frauwirth KA, Lanfranco AR, Braunstein I, Kobayashi SV, et al. CTLA-4 and PD-1 receptors inhibit T-cell activation by distinct mechanisms. Molecular and Cellular Biology. Am Soc Microbiol 2005;25(21):9543−53.

[25] Wherry EJ. T cell exhaustion. Nature Immunology. Nature Publishing Group 2011;12(6):492−9.

[26] Hino R, Kabashima K, Kato Y, Yagi H, Nakamura M, Honjo T, et al. Tumor cell expression of programmed cell death-1 ligand 1 is a prognostic factor for malignant melanoma. Cancer. Wiley Online Library 2010;116(7):1757−66.

[27] Keir ME, Butte MJ, Freeman GJ, Sharpe AH. PD-1 and its ligands in tolerance and immunity. Annual Review of Immunology 2008;26:677–704.
[28] Chang C-H, Qiu J, O'Sullivan D, Buck MD, Noguchi T, Curtis JD, et al. Metabolic competition in the tumor microenvironment is a driver of cancer progression. Cell. Elsevier 2015;162(6):1229–41.

FURTHER READING

[1] Naumov GN, Akslen LA, Folkman J. Role of angiogenesis in human tumor dormancy: animal models of the angiogenic switch. Cell Cycle. Taylor & Francis 2006;5(16):1779–87.
[2] Naumov GN, Folkman J, Straume O. Tumor dormancy due to failure of angiogenesis: role of the microenvironment. Clinical and Experimental Metastasis. Springer 2009;26(1):51–60.

CHAPTER 7

Conclusions

In this book, we started with a question: Why do metastases sometimes start growing faster (or at least appear to do so) after removal of primary tumors?

In exploring possible answers to this question, we talked about dormant tumors and mechanisms that could underlie tumor dormancy (Chapter 2). We then talked about tumor-induced immune suppression and how cancer cells can incapacitate cytotoxic lymphocytes by outcompeting them for shared nutrients (Chapter 3). We also briefly discussed how pharmacokinetic and pharmacodynamic modeling can assist in identifying therapeutic targets, and how we can simulate potential effects of therapies. In Chapter 4, we moved to the discussion of the process of normal and pathological blood vessel formation, and how small but consistent changes in the microenvironment can turn the process of normal wound healing into pathological angiogenesis that we see in tumors. In Chapter 5, we discussed the notion homeostasis of angiogenesis regulators in the body, until finally arriving at a possible answer to our question: removal of primary tumor can cause decrease in the total level of inhibitors in the body, creating a window of opportunity for the secondary tumor. Finally, in Chapter 6, we talked about possible approaches for prevention of metastatic resurgence, which converge on one important message: to target a system such as cancer, we need to target not only cancer cells themselves but also tumor microenvironment.

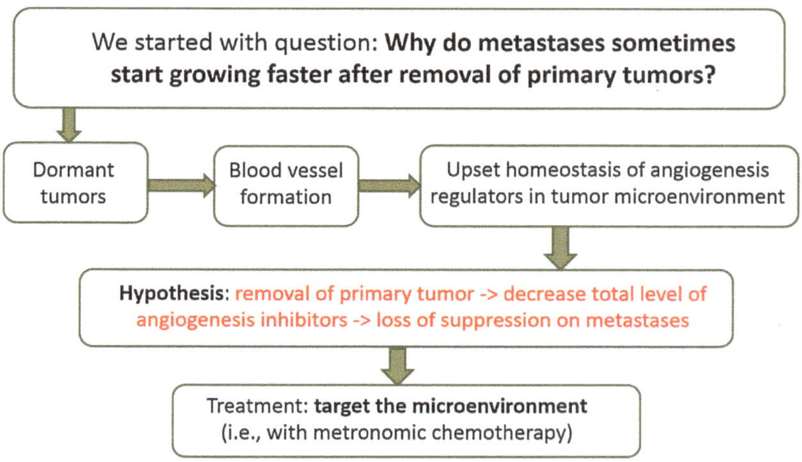

Understanding Cancer From a Systems Biology Point of View
ISBN 978-0-12-813673-7
http://dx.doi.org/10.1016/B978-0-12-813673-7.00007-0

Another goal of this book is to highlight numerous possibilities that mathematical modeling can provide for answering biological questions. Writing a model forces us to formalize what we know and what we do not know, which is not a trivial task. Modeling also requires us to prioritize different components of a biological system in an attempt to identify which component has the greatest impact on the system at each particular time. Analyzing a model allows testing hypotheses, refining our understanding of the biology and hopefully advancing it.

It is my hope that mathematical modeling will eventually become a common tool in answering biological questions as any laboratory-based technique, and together we will be able to move scientific knowledge forward.

Acknowledgments

In September 2016, I received an invitation to give a lecture at Rochester Institute of Technology, College of Science and the Thomas H. Gosnell School of Life Sciences, on a topic related to my research on cancer and mathematical modeling. As I was preparing the lecture, I realized that much of the material fit into a story, which may be of interest to others as well, a realization that was confirmed by the wonderful reception of the lecture at RIT. Therefore, that visit gave rise to this book, and I would thus very much like to thank Larry Buckley, Andre Hudson, Kate Wright, and all the wonderful faculty and students of RIT for indirectly playing a part in the creation of this manuscript.

The research that went into this book was conducted over the years 2012–16, as I entered my time as a postdoctoral fellow first at Center of Cancer Systems Biology at St. Elizabeth's Medical Center in Boston with Philip Hahnfeldt and Lynn Hlatky, and then in the lab of Giannoula Klement at Tufts Medical Center, where I learned so much about cancer dormancy and angiogenesis. I would like to thank them and my colleagues and friends Heiko Enderling, Kathleen Wilkie, Abdo Abou-Slaybi, Olga Dashevsky, among many others, for discussions, and coffees, and gatherings, and all those things that go into the creative scientific process.

But of course, my postdoctoral fellowship in Boston would never have happened if I had not been given an opportunity to do my doctoral dissertation in the most encouraging and intellectually stimulating environment at Arizona State University, where I got my PhD in Applied Mathematics for Life and Social Sciences under the supervision of Dr. Carlos Castillo-Chavez and John Nagy, to whom I am forever grateful for teaching me to be a scientist.

And all of that would not have happened without my parents, who also are my best collaborators, Faina Berezovsky and Georgy Karev. Among other things, they taught me how to write and analyze mathematical models and are always there for new scientific discussions and projects. Yes, mathematical modeling is our family business.

Finally, I would like to thank my husband Gideon Coltof, forever supportive, encouraging, patient, and kind. This book would not have happened without him.

Supplementary Chapter I: Code, Exercises, and Future Directions

In supplementary materials, two Matlab code files have been attached.

One of them describes the distributed logistic model, with equations given in Chapter 2. This piece of code allows replicating the results and of course is amenable for modification.

EXERCISE 1

1. Run the code as it is. Then try changing parameters of initial distribution μ. How does it affect time to escape from tumor dormancy? What if you change the initial population size N_0? What does it do to changes in population size, expected value, and variance as they change over time? What about the distribution of clones $P_c(t)$?
2. In this example, we used truncated exponential distribution on the interval [0 1]. What happens if you change the boundaries of the distribution, i.e., to [0 2]? What about [0 10]?
3. Evaluate the change in dynamical behavior of the key characteristics of the model if the initial distribution is different, for instance
   a. Uniform
   b. Normal
   c. Gaussian

   This will require finding the appropriate forms of the moment generating functions and calculating the expected value and the variance according to the formulas given in Chapter 2.

   How does changing the initial distribution change predictions? How does it affect time to escape from dormancy? Try different intervals for the initial distributions, similar to the original example, where the initial truncated exponential distribution was bound on the interval [0 1]. How does it change predictions?
4. What other situations can we observe these behaviors in? One example is demography by G. Karev; what others? What are the implications?

EXERCISE 2

The second file allows reproducing the model from Chapter 3, which describes the dynamics of aerobic and glycolytic tumor cells competing for shared resources with immune cells. There is also a treatment option that is added in the code: drug concentrations A_{iv} and A_p, which stand for IV drug concentration and concentration of the drug in the peripheral tissues. In this particular example, the simulated drug affects the ability of glycolytic cancer cells to absorb glucose. The parameters were chosen arbitrarily, for the purposes of illustrating the method and not for any practical predictions.

1. Look at how the effect of the drug on system dynamics compared to no drug. The effect of the drug is implemented here using the following simple functional form: $\frac{1}{\xi A_p+1}$. What are the effects of the drug if $A_p = 0$? What about when $A_p > 0$? What is the effect of parameter x_i?
2. Try different protocols: different dose numbers, different dose levels, different intervals between doses. What do you observe?
3. In the text, we discussed various other parameters that can be affected by the drug. Can you implement the effect of treatment on immune cell activation (i.e., parameter i_0)? What about immune cell mortality (parameter μ_I) or cancer cell mortality (parameter μ_T)? Note: While the idea is the same, if the drug affects cell mortality, the functional form needs to be modified.
4. Play with different protocols.
   - **a.** Can you find a method to optimize predictions for different tumor types, with different growth rates, death rates, immunogenicity?
   - **b.** What about various degrees of sensitivity to treatment?
   - **c.** A very interesting case is trade-offs for chemotherapy, which kills tumor cells, and with regards to immune cells, can both stimulate and harm the immune cells. Can you find a way to optimize the treatment for individual cases?
   - **d.** What about if the cancer mutates, and the only way for mutated therapy-resistant cells to be eliminated is using the immune system. Can we find an optimization technique to help immune system target therapy-resistant tumor cells?
   - **e.** What about protocol duration? Two months? Two years? Lifetime?

Code example for Exercise 1.

```
function suppl_logistic_dist_trexp
% Sample code that allows reproducing the dynamics of a
heterogeneous
% population that grows according to logistic growth law,
with Malthusian
% parameter c distributed according to truncated
exponential distribution
% on the interval [a,b]
close all
clear all
clc
options=odeset('RelTol',1.0e-10);


N0=.001; %initial population size
tmax=300; %simulation runs until time tmax
K=1; %carrying capacity
a=0; b=1; %boundaries of truncated initial distribution

mu=150;%parameter of the initial distribtion

params=[K;mu;N0;a;b]; %parameter vector
time=[0;tmax]; %time interval over which we integrate
in_cond=[0 0]; %initial conditions q(t)=0, t=0;

[T, Y]=ode23s(@dist_logistic,time,in_cond,options,params);
%ode23 is a solver that will numerically integrate function
dist_logisitc
%over time period "time" with initial conditions "in_cond",
with precision
%defined in "options" and parameter values defined in
"params"
% The results will be given in vector Y

q=Y(:,1); %auxiliary "keystone" variable from the solution
tt=Y(:,2); %time from the solution

%%% in this section we calculate moment generating
function, the expected
% value of parameter c and its variance as they change over
time
```

```
% here, we formulas are written for truncated initial
distribution, and
% should be modified for other distributions
for i=1:length(Y)
    m01(i)=(mu*(exp(a*(q(i)-mu))-exp(b*(q(i)-mu))))./((mu-
q(i))*(exp(-mu*a)-exp(-mu*b))); %moment generating function
for truncated exponential distribution on interval [0 1]
    exp_val(i)=b+1/(mu-q(i))+(a-
b)*exp(b*mu+a*q(i))./(exp(b*mu+a*q(i))-exp(a*mu+b*q(i)));
%expected value of c for truncated exponential distribution
on interval [0 1]
    variance(i)=1./(mu-q(i))^2-(((a-
b)^2)*exp((a+b)*(mu+q(i))))./((exp(b*mu+a*q(i))-
exp(a*mu+b*q(i))).^2); %variance of c for truncated
exponential distribution on interval [0 1]
    Nt(i)=N0*m01(i);% total population size
end


%%% here, we calclate the change in distribution over
clones within the
%%% population over time
mu1=linspace(0,1,10); % defining the space for paramter mu
on interval [0 1]
for i=1:length(Y)
    for j=1:length(mu1)
        p0al(i)=mu*exp(-mu*mu1(j))/(1-exp(-mu));
        pdist(i,j)=p0al(i)*(exp(mu1(j)*(q(i))))/m01(i);
    end
end


%%% plotting the results

figure
subplot(2,2,1)
plot(T,Nt,'LineWidth',2)
ylabel('N(t)')
xlabel('time')
hold on

subplot(2,2,2)
plot(T,exp_val,'LineWidth',2)
ylabel('E^t[c]')
```

```
xlabel('time')
hold on


subplot(2,2,3)
plot(T,variance,'LineWidth',2)
ylabel('Var^t[c]')
xlabel('time')
hold on

subplot(2,2,4)
surf(mu1,T,pdist,'EdgeAlpha',0)  % The 3D Grpph
ylabel('Time t'), xlabel('c'), zlabel('P^t[c]')
ylim([0 tmax])
grid on, shading faceted %or 'shading interp'
colormap(sqrt(hsv))  %We can modify the color matrix by
exponentiation ;-]
axis xy

%defining the equations that are used in ode23 solver
function F1=dist_logistic(t,y,params)
% unpacking the parameter vector
K=params(1);
mu=params(2);
N0=params(3);
a=params(4);
b=params(5);

q=y(1); tt=y(2); %renaming the variables, for convenience

m01=(mu*(exp(a*(q-mu))-exp(b*(q-mu))))/((mu-q)*(exp(-mu*a)-
exp(-mu*b))); %calculating the moment generation function
nt=N0*m01; %calculating total population size

% these are the equations that are actually being solved
% we then calculate all the necessary characteristics in
the main program
dq=(1-nt./K);
dtt=1;

F1=[dq;dtt];
```

Code example for Exercise 2.

```
function tumor_metabolism_treat
% Sample code that allows simulating treatment on a tumor-
immune-shared
% glucose model
close all
clc
options=odeset('RelTol',1.0e-6);

T0=1;C0=T0; I0=T0; %initial conditions
Aiv0=0; %initial conditions for IV dose of the drug
Ap0=0; %initial conditions for drug concentration in plasma

%define parameters for the tumor-immune equations
alpha=0.1; bt=.9; bi=bt; s=.7;
ga=0.1; gg=2.2*ga; dar=0.01; dgly=3*dar;
da=.1; dg=2*da; di=dg; g0=1; p1=.01;
e1=0.1; e2=e1-0.00;
ii0=0.0002; m=.01; ki=.01;

Ta0=(1-alpha)*T0; Tg0=alpha*T0; %proportion of glycolytic
and aerobic cells in total tumor population
Tmax=3600; %time span

%define parameters pertaining to drug administrtion and
elimination rates
k01=0.5;%drug administration rate
k10=0.001;%drug elimination rate
params=[gg,ga,da,dg,e1,di,m,ki,dar,dgly,g0,p1,alpha,ii0,e2,
bt,bi,s,k01,k10]; %parameter vector

%%%%treatment
options=odeset('RelTol',1.0e-3);

dosenum=5; %simulated number of doses
ivdose=100;%sample dose
intvl=70;% interval at which the doses are administered
cDose(1)=0;
        for i=1:dosenum
            tDose(i)=intvl*(i-1); %dosing intervals
```

```
            cDose(i)=ivdose; %update the dose at each
dosing interval
        end
ti = [tDose max(Tmax)];
%we have no created a vector that contains doses, and time
when they were
%administered
tPoints = []; cPoints = [];
%now we simulate the actual treatment: integrate the system
but at each
%dosing time point update the value of treatment
        for i = 1:length(tDose)
            Aiv0 = Aiv0 + cDose(i);
            tSpan = [ti(i) ti(i+1)];
            [tOut,cOut]=ode23s(@tumorimmunetreat,tSpan,[Ta0
Tg0 C0 I0 Aiv0 Ap0],options,params);
            tPoints = [tPoints; tOut];
            cPoints = [cPoints; cOut];
            Ta0 = cOut(end,1); Tg0 = cOut(end,2); C0 =
cOut(end,3);...
            I0 = cOut(end,4); Aiv0 = cOut(end,5); Ap0 =
cOut(end,6);% IgG0=cOut(end,7);
        end


X=cPoints;
Time=tPoints;

%%%%% rename output variables for convenience
aerobic=X(:,1); glycolytic=X(:,2); glucose=X(:,3);
immune=X(:,4);
Aiv=X(:,5); Ap=X(:,6);


%%%plot the results
figure
subplot(3,1,1)
plot(Time,aerobic,'g:','LineWidth',2)
hold on
plot(Time,glycolytic,'r--','LineWidth',2)
plot(Time,aerobic+glycolytic,'LineWidth',1)
legend('Aerobic','Glycolytic','Total')
ylim([0 max(aerobic+glycolytic)])
xlabel('time'); ylabel('tumor');
```

```
subplot(3,1,2)
plot(Time,glucose,'LineWidth',2)
xlabel('time'); ylabel('glucose');

subplot(3,1,3)
plot(Time,immune,'LineWidth',2)
ylim([0 max(immune)])
xlabel('time'); ylabel('immune');

figure
subplot(2,1,1)
plot(Time,Aiv)
xlabel('time'); ylabel('Aiv');
subplot(2,1,2)
plot(Time,Ap)
xlabel('time'); ylabel('Ap');
%


function dy=tumorimmunetreat(t,X,params)
%unpack the parameter vector
gg=params(1);
ga=params(2);
da=params(3);
dg=params(4);
e1=params(5);
di=params(6);
m=params(7);
ki=params(8);
dar=params(9);
dgly=params(10);
g0=params(11);
p1=params(12);
alpha=params(13);
ii0=params(14);
e2=params(15);
bt=params(16);
bi=params(17);
s=params(18);
k01=params(19);
k10=params(20);
%rename the variables for convenience
Ta=X(1);Tg=X(2);C=X(3);I=X(4);Aiv=X(5);Ap=X(6);

%define the equations
dTa=(ga)*C*Ta/(1+bt*C)-(e1)*Ta.*I/(I+s*(Ta+Tg))-dar*Ta;%
dTg=(gg)*C*Tg/(1+C)-e2*Tg.*I/(I+s*(Ta+Tg))-dgly*Tg;%
dC=g0-p1*C-da*Ta*C/(1+bt*C)-dg*Tg*C/(1+bt*C)*(1/.9*Ap+1)-
di*I*C/(1+bi*C);% here treatment is targeting glucose
absorption by the tumor (parameter dg)
dI=-
m*I+ii0*(Ta+Tg)+ki*(Ta+Tg)*I^2*(C/(1+bi*C))/(s*(Ta+Tg)+I);
dAiv=-k01*Aiv;
dAp=k01*Aiv-k10*Ap;
dy=[dTa;dTg;dC;dI;dAiv;dAp];
```

Glossary

Allee-type model: Population growth model that incorporates existence of a viability threshold. If the population size falls below this threshold, it tends to zero; if it is above the threshold, the population behaves like a logistic model.

Angiogenesis: The process of new blood vessel formation.

Angiogenesis regulators: Growth factors and cytokines involved in formation of new blood vessels. These include vascular endothelial growth factor (VEGF), basic fibroblast growth factor (bFGF), thrombospondin 1 (TSP-1), platelet factor 4 (PF-4), among others. Angiogenesis regulators can act as stimulators or inhibitors of blood vessel formation.

Carrying capacity: Maximal quantity that can be indefinitely supported by the environment. Carrying capacity can be fixed or dynamic, i.e., increase or decrease as a result of dynamics of the other variables in the system.

Checkpoint inhibitors: Drugs that target mechanisms of immune cell activation and survival (checkpoints). Normally, checkpoints act as protection against autoimmune diseases. Blocking checkpoints allows immune cells to become and stay activated longer, increasing tumor cell kill by the immune system.

Cytotoxic lymphocytes: Immune cells that are capable of killing cancer cells.

Heparan sulfate (HS): A component of the extracellular matrix, a polysaccharide that among its other functions, is critical for successful signaling of angiogenesis regulators during neovascularization.

Logistic growth: Growth that results in the population increasing until it reaches its carrying capacity.

Malthusian growth: Growth law that predicts the population reaching infinite size in finite time. The term is named after Thomas Robert Malthus, who described potential resource depletion problems that an exponentially growing population can encounter in his 1798 essay on population growth.

Mouse xenografts: Implanting human-derived cancerous tissue into mice that are often immune deficient.

Metronomic chemotherapy: Low-dose high-frequency administration of cytotoxic drugs (administered regularly, like a metronome, hence the term).

MTD: Maximum tolerated dose, the standard protocol for chemotherapy administration. It involves administering highest doses of cytotoxic drugs that a patient can tolerate and requires prolonged periods of time between treatments to allow the patient to recover. Since cancer cells are not equally susceptible to chemotherapeutic drugs, this approach can result in killing therapy-sensitive cells, leaving behind therapeutically resistant cancer cells.

Ordinary differential equations (ODEs): Equations that describe the rate of change of a variable or variables (i.e., population size) with respect to time. The term ordinary is used in contrast to partial differential equations, which track the changes over time of more than one independent variable (i.e., time and space, or age).

PKPD: Pharmacokinetics and pharmacodynamics. Pharmacokinetics describes the dynamics of the drug as it goes through absorption, distribution, metabolism, and eventually excretion. Pharmacodynamics describes the effect the drug has on its target when it is present in the body.

Platelets: Microparticles produced by megakaryocytes in the bone marrow, which circulate in the body, store angiogenesis regulators, and are involved in clotting and neovascularization.

Sensitivity analysis: Technique used to determine how changes to parameter values impact model predictions; it is used to identify parameters that have the greatest impact on the predicted dynamics.

Tumor dormancy: A phenomenon when the tumor is present but is not growing. It has also been termed "cancer without disease." Primary tumor dormancy refers to a dormant state of tumors in a preclinical stage. Secondary or metastatic tumor dormancy refers to dormancy of tumors after the growth of primary tumors.

INDEX

'*Note*: Page numbers followed by "f" indicate figures, "t" indicate tables.'

D

E

G

N

P

R

S

T

U

V

W

FSC
www.fsc.org
MIX
Papier | Fördert
gute Waldnutzung
FSC® C083411

Kolibri 360 GmbH
Ferdinand-Jühlke-Straße 7
99095 Erfurt, Deutschland
produktsicherheit@kolibri360.de